Advanced Series in Agricultural Sciences 21

Advanced Series in Agricultural Sciences

Volumes already published in the series

Volume 1: *A. P. A. Vink*
Land Use in Advancing Agriculture

Volume 2: *H. Wheeler*
Plant Pathogenesis

Volume 3: *R. A. Robinson*
Plant Pathosystems

Volume 4: *H. C. Coppel, J. W. Mertins*
Biological Insect Pest Suppression

Volume 5: *J. J. Hanan, W. D. Holley, K. L. Goldsberry*
Greenhouse Management

Volume 6: *J. E. Vanderplank*
Genetic and Molecular Basis of Plant Pathogenesis

Volume 7: *J. K. Matsushima*
Feeding Beef Cattle

Volume 8: *R. J. Hanks, G. L. Ashcroft*
Applied Soil Physics

Volume 9: *J. Palti*
Cultural Practices and Infectious Crop Diseases

Volume 10: *E. Bresler, B. L. McNeal, D. L. Carter*
Saline and Sodic Soils

Volume 11: *J. R. Parks*
A Theory of Feeding and Growth of Animals

Volume 12: *J. Hagin, B. Tucker*
Fertilization of Dryland and Irrigated Soils

Volume 13: *A. J. Koolen, H. Kuipers*
Agricultural Soil Mechanics

Volume 14: *G. Stanhill*
Energy and Agriculture

Volume 15: *E. A. Curl, B. Truelove*
The Rhizosphere

Volume 16: *D. P. Doolittle*
Population Genetics: Basic Principles

Volume 17: *A. Feigin, I. Ravina, J. Shalhevet*
Irrigation with Treated Sewage Effluent

Volume 18: *D. Gianola, K. Hammond*
Advances in Statistical Methods for Genetic Improvement of Livestock

Volume 19: *I. Rosenthal*
Electromagnetic Radiations in Food Science

Volume 20: *D. Russo, G. Dagan*
Water Flow and Solute Transport in Soils

Volume 21: *G. Hunsigi*
Production of Sugarcane: Theory and Practice

G. Hunsigi

Production of Sugarcane

Theory and Practice

With a Foreword by A.G. Alexander

With 51 Figures and 93 Tables

Springer-Verlag
Berlin Heidelberg New York
London Paris Tokyo
Hong Kong Barcelona
Budapest

Prof. Gururaj Hunsigi, Ph.D.
Head of Department of Agronomy
The University of Agricultural Sciences, GKVK
Bangalore 560 065, India

ISBN 3-540-56552-3 Springer-Verlag Berlin Heidelberg New York
ISBN 0-387-56552-3 Springer-Verlag New York Berlin Heidelberg

Library of Congress Cataloging-in-Publication Data. Hunsigi, Gururaj. Production of sugarcane: theory and practice/Gururaj Hunsigi. p. cm. -- (Advanced series in agricultural sciences; 21) Includes bibliographical references (p.) and index. ISBN 3-540-56552-3 (acid-free paper: Berlin). -- ISBN 0-387-56552-3 (acid-free paper: New York) 1. Sugarcane. I. Title. II. Series. SB231.H865 1993 633.6' 1 -- dc20 93-14008

Printed in Germany

Typesetting: Macmillan India Ltd., Bangalore 25

31/3145/SPS-543210 – Printed on acid-free paper

Affectionately dedicated to
my father, Seshacharya, wife, Indumathi and
children, Rajashree, Prahalad and Purushottam

Foreword

For planters of sugarcane, the late twentieth century can be viewed as the best of times and the worst of times. Yet, whatever the planter's view-point and whatever the circumstances of his local sugar industry, the new book by Dr. Gururaj Hunsigi, *Production Of Sugarcane: Theory And Practice*, will be seen as a perceptive and useful contribution to his field. Moreover, it is a most timely contribution regarding the survival of sugarcane at a crossroad in its long history as a cultivated crop plant.

The "time of troubles" for sugarcane planters is most evident in developed countries but is not limited to them. Rising production costs, alternative sweeteners, the economics and politics of sugar marketing, and issues of concerned environmentalism are clouding the outlook for sugarcane planting as it was known in the past. In some regions the diversification of sugarcane to alternative farm commodities seems appropriate. Other views favour an internal diversification of sugarcane to new products and use of the whole cane plant. Fortunately, Dr. Hunsigi's work never loses sight of sugar as a permanently decisive product, for sugar has been and will always remain involved in the destiny of most nations of the sugarcane world.

Very competent books and monographs on sugarcane production have been published in the past. Ironically, some excellent works originated in countries that today are most likely to lose sugarcane as a future agricultural resource. What most clearly sets *Production* apart is its in-depth treatment of the basics of sugar planting while at the same time identifying new avenues for change and diversification.

There is little need to dwell on the continuing importance of cane sugar production in world agricultural affairs. Quite aside from the long-secure role in sweetening and fermentation processes, the potential of cane sugar as a calorific nutritional resource alone would justify its permanent production in a densely populated and underfed future world. No other plant can convert so much radiant solar energy into so much useful calorific energy in behalf of so many people. *Production* very nicely gives leeway to such perceptions.

Equally important, but in a different context, Dr. Hunsigi's *Production* offers a more valid demographic basis for future sugarcane agriculture than any previous book brought to this reviewer's attention. In contrast to so many absentee-owner industries of the nineteenth and early twentieth centuries, which at best exploited cheap labor and land in tropical climates, Dr. Hunsigi's work accommodates growers who like to plant cane, who want to plant cane, and are committed to cane planting for whatever purpose. This book as a whole is a convincing point of departure for the evolution of the role of sugarcane and newly perceived values as an international natural resource.

Dr. Gururaj Hunsigi's combination of *theory* with the practicalities of production is a farsighted approach, but one grounded solidly in over three decades of fundamental *Saccharum* research in different sugar-planting countries. Enormous advances have been made in understanding sugarcane photosynthesis, sugar movement and storage, the physiology of growth and sugar utilization, and plant behavior in production regimes of extreme diversity and intensity. Few crop plants have received such scrutiny. It is correct to say that the understanding of sugarcane potentials at fundamental and tactical levels has materially outstripped understanding at conceptual program and policy-making levels. Dr. Hunsigi's *Production* makes allowances for the eventual closure of this gap in the perception of sugarcane. In any case, a serious policy blunder being made in a few "developed" regions – that of dismissing sugarcane from further affairs when sugar alone is no longer economic – should not be repeated by policy-makers who read and understand this book.

Timeliness is another attractive feature of *Production* and its salient messages. The traditional sugarcane planter is being confronted by a wholly new, pervasive, and permanent challenge to the ways in which he conducts his affairs. This challenge is conveniently depicted as "environmentalism", a worldwide awakening to a deteriorating planet, to man's need to coexist with his environment and the need to conserve his biosphere in behalf of all future generations. The strength and finality of this movement must not be taken lightly. It is based mostly on the actual trends toward earth and on the valid countermeasures undertaken by committed people. It is within this context that "the best of times" emerge for those who plant sugarcane, and who can also perceive its full potential in an earth-conscious future agriculture.

This crossroad is like no other in the history of sugarcane. The sugarcane industry must accept an agricultural mentor and leadership role in the environmental movement, or be dragged along by it. This is a crossroad not only for the sugarcane plant, but also for education and policy-making. The plant itself has dealt unilaterally and effectively with nature and the environment for many millions of years. The planter, not the plant, is to be tested soon by environmentalism.

Fortunately, just as *Production* never loses sight of sugar as its primary subject, it is equally true that sugarcane's great untapped potential as a botanic entity is retained in view. This is a feature of Dr. Hunsigi's work that future analysts of sugarcane history will not fail to record. As a complete and accomplished physiological system, a sugar system, an energy system, and an environmental system, sugarcane's ultimate role in behalf of mankind is only just beginning. Today, any country is fortunate when it has the tactical resources to produce sugarcane. And doubly blessed indeed are those who have the vision to utilize sugarcane to its full potential in their future agricultural programs.

Finally, Dr. Gururaj Hunsigi quite diplomatically treats several centuries of sugarcane's recent past as a plantation sugar crop. However, European interests first encountered sugarcane, quite unexpectedly, over 23 centuries ago in the Indus Valley of northern India. Alexander's eastern expedition observed and described it there as a closely managed commodity long familiar to local growers. What a splendid touch of fate it is that such an encouraging message regarding sugarcane production should emerge today from India, in the worst of times and best of times for sugarcane.

Río Piedras, Puerto Rico
June 1993

Alex G. Alexander

Preface

It is presumptuous on my part that I should suppose that I could unravel the finest botanical, physiological and agronomic aspects of this wonderful crop – sugarcane. However, with the ending of Earth's summit, the cri de coeur is to save the planet from environmental pollution and sugarcane is well suited as an energy crop in agroforestry or in the farm forestry system to help contain pollution due to greater amounts of CO_2 fixation. As an energy crop sugarcane may limit the denudation of forests. In its more than 300 years history sugarcane has travelled first from India to Spain, thence to the New World and has gradually spread to the Caribbean and to South and North America. Sugarcane was recognized as only a monolithic crop, i.e. producing sugar, but it is increasingly becoming a multi-product commodity, i.e. providing food, fuel, fibre and fertilizer.

This book is presented in two parts with a total of 12 chapters. The first part deals with origin, distribution, soil and climatic requirements, seed bed preparation, cultural and nutrient requirements, fertilization, irrigation and drainage, ratooning, weeds, pests, diseases, ripening and harvest. The second part deals with energy and fibre cane, cane development, technology transfer and improved aspects of sugar manufacture and byproducts. Treatment of the subject is global as the practices of nutrient and water management etc. are drawn from Asia, Africa, Oceania and Latin America. It may serve as a textbook for graduate and postgraduate students and as a reference book for personnel working in factories for sugar – paper complexes and for the general elite interested in the tropical sugar crop. In essence, this book should serve as vade mecum for cane growers, sugar and sugarcane technologists, graduate and postgraduate students, and teachers.

My debt to other scholars is great and involves too many names to list. My gratitude goes to Dr. A. G. Alexander of Puerto Rico, who has been a great source of inspiration, and to Drs. P. K. Moberly, G. J. Gascho, Brian Egan, Don J. Heinz, Edward S. Lipinsky, J. M. Lynch, J. E. Atchison, Alan Wild, D. Payne and M. Leopairote who have permitted me to use their material and/or provided useful suggestions. I owe much to Drs. K. V. Bhagyalaxmi and K. Mohan Naidu of the Sugarcane Breeding

Institute, Coimbatore, for their critical comments and suggestions. My special thanks go to G. Blaak of the FAO, to my mentor Mydur Anand, to M. P. Sharma of IISR, Lucknow, C. A. Viraktamath of UAS and to the sugarcane experts, J. Tuljaram Rao, S. C. Srivastava. S. P. Jaiswal, and S. K. Somaiya.

I hereby place on record grateful thanks to Elsevier, John Wiley & Sons, the English Language Book Society, Longman and other Indian publishers who have permitted me to use their material. I would fail in my duty if I did not also thank the International Society of Sugarcane Technologists (ISSCT), the Sugar Technologists Association of India (STA) and the South Indian Sugarcane and Sugar Technologists Association (SISSTA).

I am indebted to Dr. K. V. Devaraj, Vice-Chancellor, and Dr. G. K. Veeresh, Director of Instruction (Agri), of the University of Agricultural Sciences (UAS), Bangalore, for their constant encouragement. Thanks are also to the UAS Library staff without whose assistance I could not have completed this task. I record grateful thanks to my fellow teachers in the Agronomy Department of UAS who are the bulwark for such academic activities. I thank my friend Mr. N. Nagaraj for his assistance in compiling this work and Mr. T. R. K. Nair for neat and excellent typing of the first draft. Finally, I owe deep gratitude to my wife Indumathi who has borne with me for the past 18 months during the writing of this manuscript.

Scholars and researchers are welcome to offer their suggestions and critical comments.

Bangalore, India *Gururaj Hunsigi*
June, 1993

Contents

Part I

Part I

Chapter 1 Overture

Sugar is a highly prized commodity and has become a symbol of affluence. Nearly 60% of centrifugal sugar comes from sugarcane and the rest from sugarbeet. Sugarcane provides food, fodder, fibre and fertilizer and this world would be poorer without sugar and its by-products. The sugar industry is a major employer providing gainful employment to millions of people worldwide in the farm sector, skilled and unskilled in the processing and refining of sugar. However, tragically, the cane sugar industry was instrumental in the institution of slavery, a negation of human dignity and honour.

Commercial sugar is a disaccharide (sucrose/saccharose) composed of two monosaccharides, i.e. fructose (levulose, fruit sugar) and glucose (dextrose, grape sugar). Sucrose has the empirical formula $C_{12}H_{22}O_{11}$ and a molecular weight of 342.3 daltons. It is soluble in both water and ethanol. White sucrose crystals are monoclinic prisms having a density of 1.588 and are optically active. One teaspoonful of sugar releases 18 cal of energy. In addition, low-grade non-centrifugal sugars are consumed in Asia, including China and India, Africa and Latin America. These are called *panela* in Colombia and Venezuela, *panocha* in the Philippines, *piloncello* in Mexico, *chancaca* in Peru, *raspadura* in the Caribbean and *gur/gula/jaggery/desi* sugar in India, Bangladesh, Pakistan and Indonesia.

Interestingly, per capita sugar consumption reflects geographical and cultural differences and a consumption of over 50 kg is observed in Cuba, Brazil, the United States and Australia, while the lowest consumption (2 kg) is in Burundi and Rwanda in Central Africa. Globally, sugar consumption per head increased from 16.5 to 20 kg in 1970. During the past few years it has registered a modest increase of 2.1% but this may decline to 1–1.5% per annum by 1995. A downward trend in sugar consumption is observed in many industrialized nations of the world save the former Soviet Union and Japan. Table 1.1 suggests that the world's average sugar consumption has remained fairly static at 20 kg. Asians and Africans consume less sugar whilst the highest sugar consumption is in central North America.

It is anticipated that future years will witness an era of reduced sugar consumption, particularly in the developed world. The reasons are many: an ageing population, the cause of obesity and gallstones, a fear of becoming diabetic, having heart trouble or a variety of other diseases and, above all, a challenge from low-calorie artificial sweeteners. Even criminal behaviour is associated with sugar consumption. A general change in the life style is that 'slim is beautiful'. Sugar is invariably associated with fat and this has

Table 1.1. Per capita consumption of centrifugal sugar, continent-wise (raw value) in 1988–1989. (Anon. 1990b)

Continent	Per capita consumption ($kg\ yr^{-1}$)
World	20.8
Africa	14.1
Central North America	45.9
South America	38.5
Asia	10.5
Europe	41.8
Oceania	43.8

substantially reduced sugar consumption. One social scientist has even remarked that sugar is not the 'devil incarnate' although it is not exactly an angel either. Nutritional gurus are now convinced that except for causing dental caries, sugar consumption provides no health hazard. Hence, natural sugar from cane/beet is safe, nutritious, palatable and a cheap source of calories in a world facing food shortages, famines and starvation. Ergo, the slogan is, 'eat sugar with pleasure but with a measure'.

1.1 The Challenge from Artificial Sweeteners

With a strong desire to be slim, many affluent societies have turned to low-calorie table-top sweeteners which are aptly called 'nutritional terrorists'. These are not only expensive but have a bitter aftertaste. Nonetheless, these sweeteners have a place of prominence in the food and drink industry. Saccharin is perhaps the oldest of artificial sweeteners; it is 300 times sweeter than sucrose and leaves a bitter aftertaste. In 1982, the British Food Additives and Contaminants Committee recommended six new brands of sweeteners, namely, Aspartame, Acesulfam-K, hydrogenated glucose syrup, Isomalt, Thaumatin and Zylotol. Of these, Aspartame and Acesulfam were of commercial importance. The former is 180 times sweeter than sucrose and is made from two amino acids, while the latter is made from an intensely sweet organic salt and is 130 times sweeter than a 4% sugar solution. Acesulfam is relatively cheaper than Aspartame and these two leave no bitter aftertaste. Most sweeteners are synergistic and hence combinations, or a 'sweetner cocktail', have enjoyed the desired popularity. These sweeteners made major inroads in the industrialized nations due to advertisements and good marketing and posed a serious threat to natural sugar consumption. Nevertheless, the task force of the US Food and Drug Administration (FDA) the COMA Committee[1] of the British Government and several United Nations

[1] Committee on the Medical Aspects of Food Policy of the British Government.

agencies concluded that natural sugar from cane/beet is safe and cheap source of calories in both the developed and developing world.

The case of high fructose corn syrup (HFCS) is different since it has replaced sugar in the soft drinks and food industry. It is estimated that HFCS can gradually replace roughly 10% of the sugar market within a decade.

Despite these challenges, sugarcane will occupy a place of prominence for high biomass/fibre production with sugar as a natural by-product. Indirect evidence is that synthetics and man-made fibres have failed to replace natural fibres like cotton, silk or wool.

1.2 Sugar Agreements

Several agreements have been reached to stabilize the prices and production of sugar. The International Sugar Agreement (ISA) was reached in 1937 but was done away with at the outbreak of the Second World War in 1939. The post-war years witnessed three international marketing agreements: the Commonwealth Sugar Agreement (CSA), the US Sugar Act and the ISA. Under CSA, the exporters agreed to restrict total exports with negotiated price quotas based on average Commonwealth production costs which would be 'reasonably remunerative to the efficient producer'. The Lome Convention covers a broad range of sugar trade between the European Economic Community (EEC) and 46 countries in Africa, the Caribbean and the Pacific (ACP).

The important transformation in the sugar industry in Asia, Africa, South America and the Caribbean is a shift from export-oriented to domestic consumption. At present, 78% of sugar is for domestic consumption, 15% is sold in the 'free market' and 7% is under bilateral agreement. This transformation is due to the protectionistic policies of the EEC and the United States. There is an urgent need to moderate these policies and the ISA should provide a fair deal to both exporters and importers. It is reiterated that discussions on sugar be included in the new round of negotiations of the General Agreement on Tariffs and Trade (GATT).

1.3 Sugar and Sugarcane Production Trends

Trends in area, production and productivity of sugarcane in different continents during 1969–1971 to 1988 are presented, along with the projections for 2000 A.D., in Table 1.2. Productivity is decelerating in Africa, while it is unstable in Oceania and the developed world (Table 1.2). By the end of this century, world productivity of sugarcane is anticipated to increase marginally from the present 53.4 t ha^{-1} to 69.8 t ha^{-1}.

Table 1.2. Projections of area, production and productivity of sugarcane in the world. (FAO 1988)

Continent	Area (10^3 ha)	Production (10^3 t)	Productivity (t ha^{-1})	Area (10^3 ha)	Production (10^3 t)	Productivity (t ha^{-1})	Stability for productivity, i.e. yield/unit area (R^2)
	1969–1971 to 1988			2000 A.D.			
Africa	737	58 492	69.9	1 786	109 226	52.4	0.52
Central North America	2 918	163 590	56.0	2 825	201 667	71.4	0.47
South America	2 716	151 524	55.4	8 917	598 584	76.4	0.59
Asia	5 216	243 498	46.9	8 977	554 155	66.91	0.80
Oceania	315	23 794	75.5	518	38 428	73.4	Unstable
Developed	762	60 839	79.9	1 247	97 512	77.4	Unstable
Developing	11 196	574 547	51.6	21 777	1 398 417	69.5	0.84
World	11 961	635 386	53.4	23 016	1 495 929	69.8	0.85

Estimates for 2000 A.D. are made by employing the formula $(1 + r)^n = V_t/V_0$, where V_t is the terminal year value, V_0 is the base year value, n is the period between base year and terminal year, and V is the compounded growth rate.

By 2000 A.D. sugarcane production is anticipated to be 1496 million tons. Trends of sugar production follow a linear relation:

$$y = a + bt,$$

where t is time in years and a and b are constants.

By 2010 A.D. the projected sugar production is 96.97 ± 0.055 million tons ($R^2 = 0.96$) and is very much lower than the ones reported by others. Naidu (1989) projected a value of 141 million tons, while Kulkarni (1971) reported a value of 150 million tons which seems to be more realistic. However, these projections only hold good if there are no violent fluctuations in the international market or drastic changes in the pattern of sugar consumption.

1.4 Production Potential of Sugarcane

It is a common truism that sugarcane is the most efficient converter of solar energy to stored energy. This is due to distinct anatomical and biochemical characteristics associated with C_4 plants and known as Kranz[2] syndrome (Jones 1985). This includes, inter alia, efficient light absorption in the 'near blue' (480 nm) and far red (640–670 nm) spectral regions with twilight photosynthesis during pre-sunrise or pre-sunset periods (Alexander 1990). Furthermore, photosynthate mobility is high in the thickened and long internodes which serve as 'powerful vehicles' for sugar storage with no product repression (loc. cit.). The average photosynthetic rates of *Saccharum* sp. and the commercial hybrids range from 50–54 $mg\,dm^{-2}\,h^{-1}$. Bull (1969) reported a rate of 100 $mg\,dm^{-2}\,h^{-1}$ for Australian cane. Irvine (1975) demonstrated that high photosynthetic rates in cane are associated with greater leaf thickness, width, porosity and specific leaf weight. He further observed that cultivars with a high leaf area index (LAI) usually had higher cane and sugar yields per hectare. Sugar yield per ton of cane was associated with erect leaves. It was reasoned by Loomis and Williams (1963) that maximum crop productivity is achievable by optimum LAI, orientation of leaves and an adequate CO_2 supply. While estimating maximum crop productivity for some crops, these authors observed a photosynthetic rate of 14 μg (CH_2O) cal^{-1} at a total radiation of 500 $cal\,cm^{-2}\,day^{-1}$. At similar light intensities the photosynthetic rate for sugar cane was about 24 μg (CH_2O) cal^{-1} at a total radiation of 500 $cal\,cm^{-2}\,day^{-1}$.

The photosynthetic efficiency for a single leaf would be as high as 5% but under an array of leaves it varies from 2.5 to 3.5%. Burr (1969) observed an efficiency of 3.48% for the F 146 cultivar. More recently, Moore (1989) obtained a photosynthetic efficiency of 3.6%. These figures are much higher than for any C_3 species.

---

[2] Kranz means wreath or halo in German.

Even in dry matter production, sugarcane excels over many crop plants and ranges from $20\,g\,m^{-2}\,day^{-1}$ in South Africa to $44\,g\,m^{-2}\,day^{-1}$ in Hawaii. In tropical India dry weights were $50\,g\,m^{-2}\,day^{-1}$ or 182.5 $Mg\,ha^{-1}\,yr^{-1}$. Bull and Glasziou (1975) predicted a maximum cane yield of 280 $Mg\,ha^{-1}\,yr^{-1}$ with maximum quantum efficiency. Moore (1989) suggested a theoretical maximum cane yield of $129\,g\,m^{-2}\,day^{-1}$ or 470 $Mg\,ha^{-1}\,yr^{-1}$. Many farmers in the tropics have achieved a yield of 250 $Mg\,ha^{-1}\,yr^{-1}$.

1.5 Crop Growth Regulants and Cane/Sugar Production

Halmann (1990) stressed that the major application of these regulants is in the sugar industry. The main groups of plant hormones are: auxins, gibberellins, cytokinins, abscisins and ethylene. Vlitos (1974) suggested that the relative levels of hormones in the immature stem are:

Auxins – maximum
Gibberellins – high
Cytokinins – high
Abscisic acid (ABA) – low.

Some experiments led this author to conclude that sett treatment with indole-3-acetic acid (IAA), indole-butyric acid (IBA), tri-iodo benzoic acid (TIBA) and Cycocoel invariably improved germination and plant density, and possibly also cane and sugar yield. But these are for special applications under frost and cooler climates. Alexander (1973) reasoned that kinetin and benzyl-adenine (BA) treatment of seed pieces augmented biomass, cane and sugar yield by increasing the chloroplast enzyme ribulose bisphosphate carboxylase (Rubisco). However, treatment with ethrel causes the loss of apical dominance as evidenced by the development of side shoots, tillers, strut roots, and lalas (water shoots). This is consistent with the finding that ethylene stimulates peroxide activity in plant tissues leading to the destruction of auxins. The activity of ABA is pronounced in older tissues and hastens senescence.

1.6 Crop Models

A model may be defined as a summary of a coherent body of experimental data in a logical structure, and can be a mere hypothesis or a concept. Models are used to predict crop yield, pest-disease incidence and, more recently, the effect of climatic changes on crop production. There are three types of models: empirical, mechanistic and comprehensive. Empirical models have limited applications, while mechanistic and comprehensive models are more widely applicable. Among the empirical models, multiplication models are more realistic than additive models.

Early attempts were made to quantify yield through the growth index (GI), which is expressed as a function of the light index (LI), thermal index (TI) and moisture index (MI). Thus:

$$GI = LI.\ TI.\ MI.$$

Another approach was to quantify the association between the yield and yield attributes like plant density, and such a relationship could be hyperbolic or parabolic:

$$Y = PW.$$

where Y is the yield per unit area, P is plant density per unit area and W is the individual plant weight.

Employing a much simpler model, Sinha and Swaminathan (1979) observed that net photosynthesis as carbohydrate (Pn) is expressed as:

$$Pn = R.\ DI.\ 0.32\ g\,m^{-2},$$

where R is the photosynthetically active radiation (PAR) in the spectrum of 400–700 nm $\approx$ 222 cal cm^{-2} day^{-1}, DI is the number of days in a crop season when 100% light interception occurs, and 0.32 is a factor from Loomis and Williams (1963) who obtained 71 g carbohydrate m^{-2} day^{-1} of PAR or 222 cal cm^{-2} day^{-1}.

Dry matter (DM) production was calculated as:

$$DM = 0.65 \times Pn \quad \textit{or}$$

$$DM = 0.65 \times R \times DI \times 0.32.$$

At Mandya in tropical peninsular India,

$$PAR \approx 200\ \text{cal}\ cm^{-2}\ day^{-1}.$$

DI is taken as 305 days with 100% light interception.

$$DM = 0.65 \times 200 \times 305 \times 0.32 = 12\,688\,g\,m^{-2}\,yr^{-1},$$
$$= 127\ Mg\ ha^{-1}\ yr^{-1}.$$

This model takes into account abiotic stresses of soil and water (FSC/FWD)[3]. Thus, $DM = 0.65 \times R \times DI \times 0.32 \times FSC$ (or FWD).

An exponential model was used by Austin (1982), who observed that maximum DM yields are well short of the expected when the energy content of DM and PAR are taken into account. He even suggested that use should be made of growth regulants which retard the senescence of green leaves.

A concept of maximum yield potential was mooted by Loomis and Williams (1963). Thus, potential yield was expressed by Gallagher and Biscoe (1978) by the relation:

$$Y = f.I.e.H/C,$$

[3] FSC = reduction factor for soil quality; FWD = reduction factor for water deficit.

where I represents PAR expressed in MJ, e is the efficiency of conversion, H is harvest index, C represents $MJ\,g^{-1}$ dry matter and f is the factor for correcting root weight.

Employing the above equation, the maximum potential yield of sugarcane (cv. CO 62175) grown in sandy loam soils (Alfisols) was calculated as:

Total solar radiation	$= 500\ \text{cal cm}^{-2}\ \text{day}^{-1}$
PAR	$= 200\ \text{cal cm}^{-2}\ \text{day}^{-1}$
	$= 8.38\ \text{MJ m}^{-2}\ \text{day}^{-1}$
For a crop over 365 days	$= 3058.7\ \text{MJ yr}^{-1}$
Harvest index (H)	$= 0.60$
Energy content (C)	$= 4000\ \text{cal g}^{-1} = 0.0167\ \text{MJ g}^{-1}$
Photosynthetic efficiency (e)	$= 3.6\%$

Shoot–root ratio for cane grown in these soils was approximated at 10 and hence $f = 0.1$.

Using these values:

$$Y = 0.1 \times 3058.7 \times 3.6 \times 0.6/0.0167$$
$$= 39\ 561.47\ \text{gm}^{-2}\ \text{yr}^{-1}.$$

Thus, the maximum potential cane yield is approximated at $395\ \text{Mg ha}^{-1}\ \text{yr}^{-1}$. It is reiterated that high production potential in sugarcane, inter alia, is due to the long duration and to botanical and anatomical features of the plant.

Some promising non-arrowing canes, like KHS3296, KHS2045, B37172 etc., produced on average $30\ \text{gm}^{-2}\ \text{day}^{-1}$ or $3.58\ \text{g MJ}^{-1}$ (PAR) which compares with DM production in C_4 plants. Nitrogen application increased the DM production per unit of PAR.

1.7 Ideotypes

The term ideotype was first proposed by Donald (1968) and is consistent with cognate terms like genotype and phenotype. An ideotype is a biological model that performs and behaves in a predictable manner in a given environment. It is supposed to resist biotic and abiotic stresses and gives better yields, both in quality and in quantity. When developed as a cultivar, it is expressed as:

Y (econ)	= Y (biol) × H.
(economic yield)	= (biological yield) × (harvest index).

Outlining the objectives of ideotypes, Sedgley (1991) stated that defect removal in crops and selection for yield and quality are the primary aims. An approach towards an ideotype in sugarcane was attempted by Hunsigi and Srivastava (1975) and the relation is:

$$Y = 796.0247 + 0.7905\ X_1 + 11.5959\ X_2 + 3.9776\ X_3 \quad (R^2 = 0.79),$$

Table 1.3. Ideotypes in sugarcane. (After Moore 1989)

Ideotypes for	Some important characteristics
Better light utilization	Increased spectral absorption at 400–700 nm Increased efficiency of CO_2 reduction per quantum of light Efficient transport of CO_2 from leaf and sugar from phloem
Better light absorption	Increased (LAI) and leaf area duration (LAD), tillering More erect and spirally arranged leaves Increased mesophyll thickness Higher content of chlorophyll and carotenoids Higher stalks m^{-2}
Better CO_2 uptake	Increased stomatal conductance for CO_2 Decreased leaf boundary resistance to CO_2 Decreased leaf width
Better partitioning of photosynthates	More partitioning to leaves at an early stage More partitioning to storage at a late stage More efficient mobility of photosynthate to the storage pool
Reduced crop losses	Reduced maintenance respiration Reduced stress loss Reduced flowering Development of non-flowering canes
Better nutrient-use efficiency	Efficient absorption and utilization of N, P and K Less antagonism among secondary/minor elements[a] Avoidance of toxicity levels of minor elements, i.e. Mn, Fe, Al[a]

[a] Author's additions.

where X_1 is the stalk number, X_2 is stalk length (cm), X_3 is weight per cane (kg) and Y is yield per plot (kg).

An ideal plant type for cane yield should possess at least 30% light transmission ratio (LTR) at the grand growth phase so as to accommodate 20–30 millable canes m^{-2} at harvest. Such plant types showed a direct relationship between crop growth rate (CGR) and LAI up to an LAI value of 7. Moore (1989) has described different ideotypes in cane for better light utilization, CO_2 uptake, photosynthate partitioning etc. (Table 1.3).

Chapter 2 Introduction

Sugarcane, a sweet reed, has attracted the attention of many, from monks to monarchs. Modern sugarcane is a complex hybrid of two or more of five species of the genus *Saccharum* and belongs to the family *Poaceae*, subfamily *Panicoideae*, tribe *Andripogoneae* and subtribe *Saccharininae*. Heinz (1987) stated that this subtribe has two natural groupings, i.e. *Saccharastrae* and *Eulaliastrae*, but which are not formal taxonomic units. Presently, this giant grass is grown in an area of 16 million hectares in over 79 countries.

2.1 History

Sugarcane is intimately associated with Indian mythology. The word *Saccharum* seems to originate from the Sanskrit word *Sarkara*. It symbolizes prosperity for it adorns the goddess of wealth, Sri Laxmi (Fig. 2.1). The most ancient reference to sugarcane is in *Atharvaveda* (5000 years old) which reads, 'I have crowned thee with a shooting sugarcane so that thou shalt not be averse to me'.

Deerr (1948) has detailed the occurrence of sugarcane in legendary India. He observed that *Ikshu* means sugarcane but has no relation to *Ikshvaku*, the solar dynasty founded by Manu, the father of mankind. It is believed that sage Vishwamitra created sugarcane in a rival heaven for his disciple Trishanku, a descendant of Ikshvaku. In the Solomon Islands it is also thought that mankind originated from a cane variety *tohononu*.

Sugarcane is referred to extensively in Buddhist literature. Gautama, the Buddha, was known as the 'king of sugarcane' or was believed to have been born from sugarcane. According to Deerr (1948) the earliest reference of sugarcane bagasse being used as fuel was in around 100 A.D.

2.2 Origin and Distribution

Brandes and Sartoris (1936) state that the 'cradle of cultivated sugarcane is the region where the two wild species i.e. *Saccharum spontaneum* and *S. robustum* are found'. The authors hypothesized that species with higher chromosome numbers are confined to the tropics while those with lower

Fig. 2.1. The goddess of wealth, Sri Laxmi, with sugarcane in her hand

numbers extended into the subtropics. Through his cytological studies Parthasarathy (1948) indicated that thin, hardy canes, *S. barberi* Jesw. and *S. sinense* Roxb., originated in north India by promiscuous hybridization of *S. spontaneum* and *S. officinarum*. However, recent evidence is at variance with this. The origins of *S. barberi* are in north India but those of *S. sinense* are in China. The thin, hardy north Indian canes, *S. barberi*, named after C. A. Barber, have narrow leaves and can tolerate adverse climatic conditions. 'Chunnee' canes are represented by this group. The Chinese canes, *S. sinense*, have evolved through introgression between *Erianthus*, *Miscanthus*, *Sclerostachya* and *S. officinarum* (Heinz 1987). The Chinese canes are tall with broad leaves, represented by the 'Pansahi' clone, the typical one being the 'uba' cane. Both the Indian and Chinese canes were carried by Arab traders to Persia, Syria and the islands of Cyprus, Malta and Sicily in the Mediterranean. Alexander the Great took with him, after the conquest of India, this 'honey reed'.

The highly prized cane is *S. officinarum* L. and earlier investigators suggested that it originated in New Guinea. It is essentially a 'garden cane' and is not found in the wild. Alexander (1973) even suggested a continuous 'land bridge' between the enormous Asiatic and Australian continents during the Cretaceous period which aided the spread of canes to Melanesia. Roach and Daniels (1987) held the view that the origin of *S. officinarum* is to the east of the Wallace line (Fig. 2.2). Artschwager and Brandes (1958) and

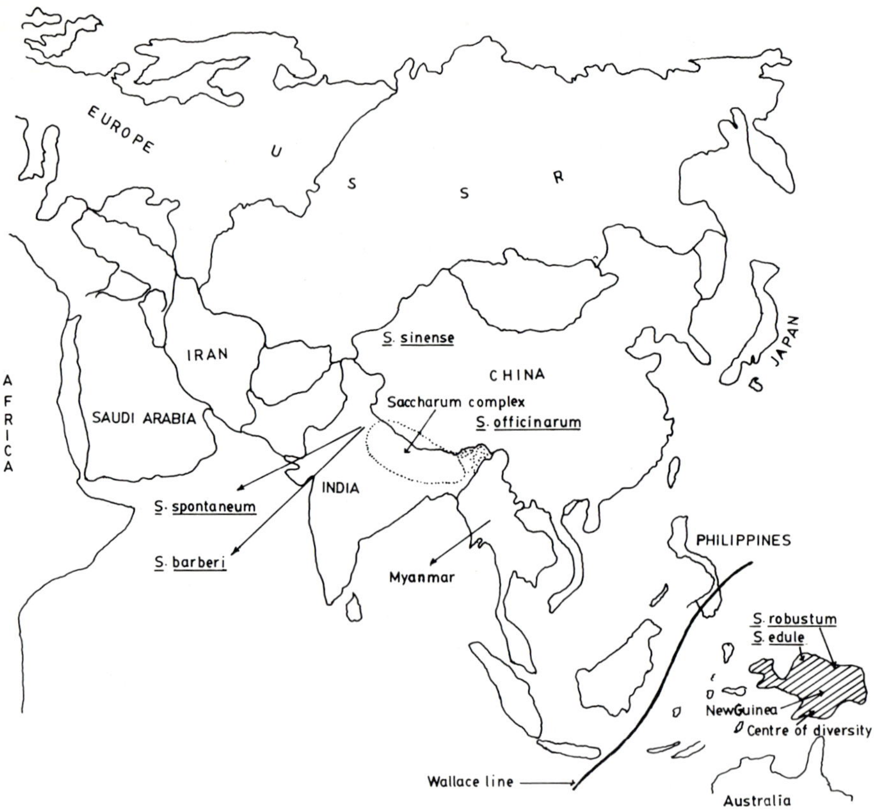

Fig. 2.2. The origin of *Saccharum* species

Grassl (1969) have put forth the suggestion that *S. robustum* is the closest wild relative and putative ancestor of *S. officinarum*. Thus, the origin of *S. officinarum* seems uncertain but Heinz (1987) concluded that it is evolved from *S. robustum* by introgression from other genera, and that the origin is from the Indo-Myanmar[1]-China border with New Guinea forming the main centre of diversity (Fig. 2.2).

Among the wild canes found in swamps and uplands, *S. spontaneum* L. has a wide distribution in Asia and Africa and has contributed substantially towards the development of modern hybrids. These are perennial grasses with aggressive rhizomatous tillering. The clones exhibit marked resistance to moisture stress, low temperatures and many pests and diseases attacking sugarcane. The centre of origin of *S. spontaneum* is the colder regions of subtropical India (Rao 1989). Another wild cane, *S. robustum* Jesw., is

[1] Burma has recently been named Myanmar.

Table 2.1. Principal characteristics and probable origin of different species of *Saccharum*

Species	Sucrose content	Fibre content	Maturity	Adaptability	Stem	Leaves	Reaction to diseases			Probable origin
							Sereh	Mosaic	Smut	
S. sinense	Medium	High	Mid-early	Tropical and subtropical	Long and slender	Long and narrow	Immune	Susceptible	Some susceptible	China
S. spontaneum	Very low	Very high	Early	Tropical and subtropical	Slender	Very narrow	Immune	Immune	Moderately susceptible	Cold regions of subtropical India
S. barberi	Medium	High	Mid-early	Tropical and subtropical	Medium and slender	Short and narrow	Immune	Susceptible but some tolerant	Moderately susceptible	North India
S. robustum	Low	Very high	Variable	Tropical	Very long and thick	Broad to medium	Undeter-mined	Susceptible	Undeter-mined	New Guinea
S. officinarum	High	Low	Variable	Tropical and subtropical	Long and thick	Long and broad	Susceptible	Susceptible	Moderately susceptible	Indo-Myanmar-China border
S. edule	Low	High	—	Tropical	Very long	Dense and thick	Pubescent, velvety	Susceptible	Moderately susceptible	New Guinea

S. robustum and *S. spontaneum* are wild species.
Some maintain that *S. edule* is not an authentic *Saccharum* sp. but a hybridization between *S. robustum* and *Miscanthus floridulus* or a simple mutant of *S. robustum.* Abortive cauliflower-like flowers are edible.
The origin of the different species of *Saccharum* is based on the present literature.
Burma renamed as Myanmar.

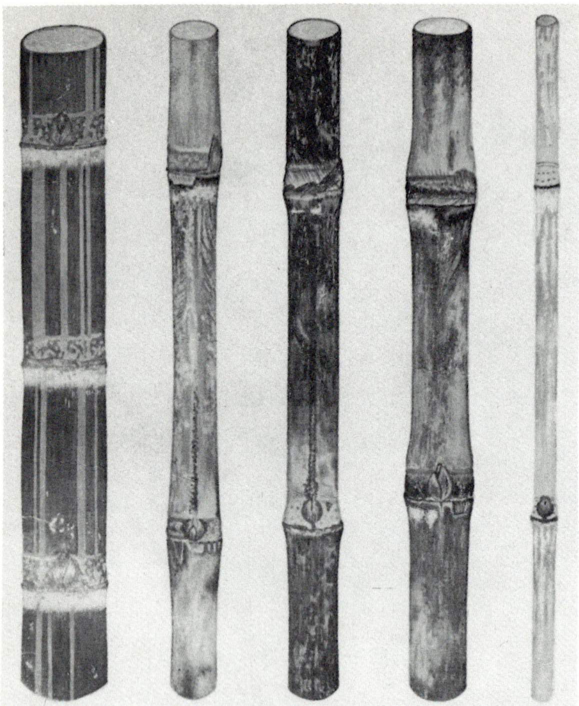

(1) (2) (3) (4) (5)

Fig. 2.3. Typical clones of *Saccharum* species 1, *S. officinarum* (noble cane); 2, *S. barberi* (Indian cane); 3, *S. sinense* (Chinese cane); 4, *S. robustum* (wild cane); 5, *S. spontaneum* (wild cane)

endemic to New Guinea and the islands of Melanesia. Its habitat is mostly river banks and water courses. The plants are tall with a high fibre and low sucrose content. It originated in New Guinea. One small group of canes, which could hardly be 'dignified by the name sugarcane' but which is closely related to *S. robustum*, is *S. edule* Hask. This may have evolved as a mutant of *S. robustum* or by introgression between *S. robustum* and *Miscanthus*. The centre of origin is New Guinea. The aborted inflorescences are used as food by the natives of Melanesia.

The origins of *Saccharum* spp. are shown in Fig. 2.2, while their general characteristics are given in Table 2.1. Typical clones are illustrated in Fig. 2.3.

2.3 The *Saccharum* Complex

Mukherjee (1957) viewed that this complex consists of four genera, namely, *Saccharum*, *Erianthus*, (section Ripidium) *Sclerostachya* and *Narenga*, to which may be added *Miscanthus* (section Diandra), *Imperata* and *Sorghum*

(Fig. 2.4). It is generally agreed that the origin of the *Saccharum* complex is the Indo-Myanmar-China border (Fig. 2.2). Mukherjee (1957) reiterated that India is the centre of maximum variation of the *Saccharum* complex with a region of concentration in the Bengal-Assam-Sikkim zone. It has been pointed out that the members of this genera have found wide use as material for building, fencing, shelters and wind breaks and as medicinal plants. Young inflorescences and shoots of *S. spontaneum* were eaten as *lalab* by the natives of New Guinea.

According to Grassl (1969), *Erianthus maximus* is considered to be the intergeneric hybrid of *S. officinarum* × *Miscanthus floridulus*. *Erianthus* has a wide but discontinuous distribution, mainly in Asia, and *E. arundinaceus* is similar to sugarcane. Furthermore, *Saccharum* is the most advanced genus, while *Sclerostachya* is the most primitive. *Sclerostachya* and *Narenga* are closely related and hence can intercross freely. Under suitable conditions, the allied genera will cross and produce viable seeds. Thus, intergeneric crosses of *S. officinarum* with *Sorghum durra*, *Bambusa arundinaceae*, *Erianthus*, *Imperata* and *Zea* have been attempted.

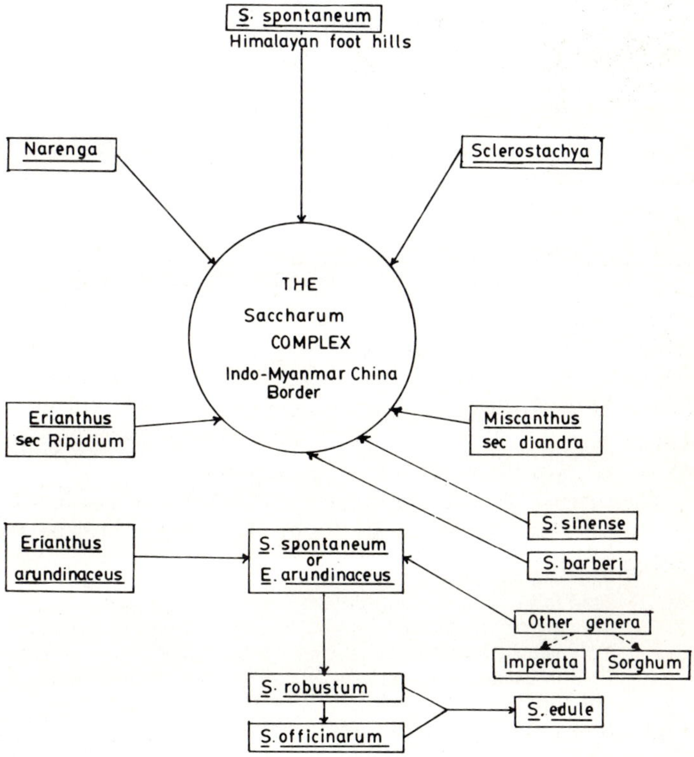

Fig. 2.4. The *Saccharum* complex

2.4 Noble Canes and Nobilization

The Dutch in Java used the term 'noble canes' for *S. officinarum* with obvious reference to the thick, flamboyant colours of these tropical species. These are well-tended 'garden canes', low in fibre, self-detrashing, high in sucrose, and used for chewing and sugar making.

The distribution is well documented. Primary migrations were to the New Hebrides and New Caledonia and there were secondary migrations into Indonesia, the Philippines, Indo-China, the Malay Peninsula and India. These garden canes accompanied man in his early travels to Fiji, Tahiti, South Asia, India and Africa. A noble cane 'Creole' was brought to the Western Hemisphere by Columbus in his second voyage in 1493. Alexander (1985) observed that the noble canes became the modern canes of commerce and reached the Western Hemisphere via Spain. Some of the important noble canes which ruled the 'sugar world' were Otaheite, Cheribon (Creole), Caledonia (Malabar), Badila, etc. However, by 1925, the noble cane era had ended due to serious diseases like mosaic, sereh, smut and red rot. This heralded an era of nobilization products. The process of nobilization in sugarcane is a modified back crossing which is defined as the crossing of the wild cane *S. spontaneum* L. with *S. officinarum* L. and with a repeat of the back crossing to the noble parent (*S. officinarum*). This ensures the ennobling of the wild species to produce thick stalks of good quality. Brandes and Sartoris (1936) contend that the nobilization process increases the chromosome complement of the hybrid and that usually a doubling of the chromosomes of *S. officinarum* takes place. However, there are exceptions. Thus, the nobilization products ushered in an era of hybrids and by 1929 the most popular hybrid, POJ2878 (POJ2364 × EK28), had been developed; this

Table 2.2. Location prefixes for different sugarcane varieties

Prefix	Country	Prefix	Country
B	Barbados	L	Louisiana, USA
CB	Campos, Brazil	L	Lyallpur, Pakistan
CP	Canal Point, Florida	M	Mauritius
NG	Collections from New Guinea	M	Mayaguez, Puerto Rico
D	Damerara, Guyana	P	Peru
F	Formosa, Taiwan	POJ	Proefstation oost Java
H	Hawaii, USA	PR	Insular Experimental Station, Riopiedras, Puerto Rico
J	Jamaica		
N	Natal, South Africa	PJ	Pintung, Taiwan
NCO	Bred in Coimbatore and selected in Natal, South Africa	Q	Queensland, Australia
		TUC	Tucuman, Argentina
CO	Coimbatore, India	NA	Santa Rosa, Argentina
COC	Bred in Coimbatore and selected in Cuddalore, India	RB	Alagoas, Rep. Brazil
		SP	Sao Paulo, Brazil (bred in Camamu, Bahia)
KHS	Karnataka, Hebbal sugarcane, India (Formerly HM, Hebbal, Mysore, India)	US	Experimental Station Clones, US Department of Agriculture

earned the sobriquet the 'wonder cane'. Some important nobilized hybrids are: CO419, CO421, NCO310, PR980, B37172, B62163, CB41-76, CP65-357, F160, etc. Some location prefixes for different varieties are given in Table 2.2.

2.5 Flowering and Its Control

The discovery, or re-discovery, in 1888 by Solt Wedel in Java and by Harrison and Bovell in Barbados (1888) that *S. officinarum* produces fertile seeds opened new vistas in cane improvement but flowering puts a ceiling on yield and quality. Hence, under commercial cultivation, flowered canes are invariably accompanied by reduced yield and quality. The disadvantages of flowering are: reduced quality, pithy canes, loss of apical dominance, development of side shoots, shorter leaf blades with lowered yield, high fibre and reduced juice extraction. A rough estimate gives a reduction of 25% in yield and 1% in sugar in a profusely flowering (80% or more) cane. Alexander (1973) recognized four distinct stages in the flowering process:

1. Initiation of primordia.
2. Floral organization.
3. Floral maturation.
4. Tassel emergence.

Coleman (1968) aptly stated that flowering in sugarcane is an intricate process and most commercial varieties are short-day plants. In other words, flowering depends more on the night length (nyctiperiod) than on the day length. But at the Equator with equal day and night lengths sugarcane tassels all the year round. Factors influencing flowering are:

1. Photoperiod.
2. Temperature.
3. Elevation.
4. Plant age.
5. Cultivar.

1. Photoperiod: This is the principal decisive factor and a critical dark period of 11.5 h is required prior to a high light intensity for at least 12 h duration. Latitude strongly affects flowering. Juang (1968) observed that as the latitude was lowered, the floral initiation became earlier. Similar observations were made by Panje and Srinivasan (1959) who concluded that every latitudinal degree away from the Equator delayed flowering by 2–4 days, and that every degree towards the Equator advanced it by the same interval.
2. Temperature: The night temperature during the critical period has a profound influence and low temperatures inhibit flowering. The minimum temperature for flower induction seems to be 15–18°C and the optimum night temperature is around 21–27°C.

3. Elevation: At higher elevations (700–800 m) sugarcane rarely flowers, but sparse flowering is observed at 300–400 m. The elevation-temperature interaction is obvious and at higher elevations low temperatures prevent flowering.
4. Age of the plant: For floral induction the plant should have passed the juvenile stage and Coleman (1968) suggests that as a rule of thumb three joints should be visible on the ground during the photo-inductive period. We have observed that cane planted in Mandya (12°N latitude) during May would be mature enough to flower in November to December of the same year (7 months), but would not flower in the same year if planted in June. It is reasoned that a sufficient 'flowering stimulus' is required before the 'ripeness to flower' stage is reached for floral initiation. Other factors, like nitrogen status, rainfall, soil moisture, variations in ambient temperature, etc., alter the flowering behaviour and perhaps these are responsible for yearly variations in the intensity of sugarcane flowering. Drought and diseases like mosaic greatly influence flowering.
5. Non-arrowing cane cultivars: Conscious efforts were made at the Regional Research Station, Mandya to select varieties which are reluctant to flower. The performance of some of these cultivars is presented in Table 2.3.

The promising non-arrowing canes in peninsular India are KHS2045 and B37172 which produce high yields and good quality. Our experience has been that B37172 is reluctant to flower anywhere.

The physiology of flowering (Fig. 2.5) has been succinctly described by Coleman (1968). He recognized five steps for tassling in mature cane. The first step in the 'ripeness to flower' stage is the exposure of the plant to high light intensity for at least 12 h followed by a critical dark period of 11.5 h (second step). During the dark period, the phytochrome probably disappears and there is the formation of the 'flowering stimulus' which is translocated to the apex (sink) through the complicated vascular system (third step). The fourth step would involve the accumulation and fixation of

Table 2.3. Performance of non-arrowing canes at the Regional Research Station, Mandya (12°N latitude); average of 3 years (1968–1970, 1971–1972)

Cultivar	Flowering (%)	Cane yield ($Mg\,ha^{-1}$)	Pol in juice[a] (%)
KHS2045	0.0	125	19.20
KHS2720	7.5	136	19.90
KHS2631	0.0	110	19.07
B37172	0.0	152	20.42
CO62175 (Check)	35.0	181	19.84
CO419 (Check)	35.0	117	20.05

[a] Pol is the polarimetrically determined sucrose content.

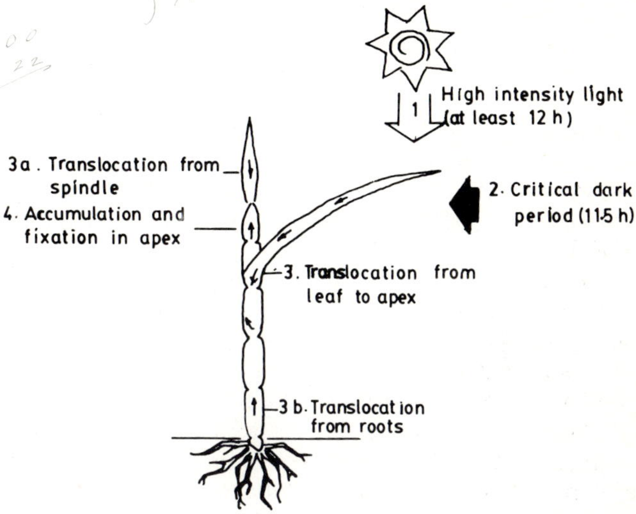

Fig. 2.5. Physiology of flowering. (Coleman 1969)

the 'stabilized stimulus' at the apex, followed by the differentiation of the floral primordia (Fig. 2.5). For a grower, flowered cane heralds a reduction in yield and quality. Several attempts have been made to control flowering and the methods include use of flares, withholding water, mechanical defoliation and use of chemical sprays at the critical stage. Withdrawing water at the 'ripeness to flower' stage in a monsoon climate is difficult to achieve. But chemical sprays, like malic hydrazide, pentachlorophenol, 6-Azauracil, paraquat, etc., have been used to inhibit flowering with variable success. The principle involved is removal of the source, i.e. leaves and spindle, at the 'ripeness to flower' stage so that the 'flowering stimulus' does not reach the sink (apex). Removal of the spindle is important for it is most active in producing the 'stimulus'. Even trimming the leaves and spindle has the desired effect. This author confirms that defoliation, including removal of the spindle, by one spray with paraquat (1,1'-dimethyl-4,4-bipyridylium chloride) at 3–5 $l\,ha^{-1}$ dissolved in 900–1000 $l\,ha^{-1}$ of water (high-volume spray) during the 'ripeness to flower' stage, i.e. during the first 2 weeks in August, at Mandya (12°N latitude) completely inhibited flowering in profusely tassling (35–40%) varieties like CO419 and CO62175 with a consequent improvement in yield and quality. In Brazil, ethephon sprayed in February (cv. NA56-79 and SP70-1143) completely controlled the tassels and increased both sugar and alcohol production. This chemical also showed promise for increasing the yield by suppressing flowering and pithiness (Silva et al. 1989). It is also claimed that ethephon can reduce the clarification and processing time in the factory (Coleti et al. 1986).

The bottom line is that Alexander (1973) prophesied that the future grower will have access to new control measures which are less drastic than chemical defoliants and drought treatments.

2.6 The Advent of Short Duration/Early Maturing Rich Cane Cultivars

Research was undertaken to introduce short duration/early maturing canes to fit into the multiple cropping system and to effectively utilize resources like land, water and other inputs to augment sugar production per unit time. The strategy was to produce three crops (one plant and two ratoons) in 24 months.

The importance of early maturing canes was realized even in the 1930s and sugarcane, sorghum and maize crosses were attempted with the sole objective of developing early maturing canes. Thus, in the 1980s great strides were achieved by the Sugarcane Breeding Institute at Coimbatore in India and early maturing rich canes were developed. These cultivars mature in 10 months with 18–20% pol and a production potential of 9–10 $t\,ha^{-1}\,month^{-1}$ *or* $\approx 1\ t\,ha^{-1}\,month^{-1}$ sugar. Mention may be made of varieties like COC671, CO7704, COJ64, COA7601, KHS3296, etc. with the most promising being COC671 (Q63 × CO775) in tropical India and COJ64 (CO976 × CO617) in subtropical India. The production potential of some early and mid-late varieties is given in Table 2.4 (Srinivasan 1989). It is obvious that the early group have a low yield potential due to a reduced growth period; they tended to accumulate more sugar compared with dry matter. Nevertheless, the yield levels of both groups are on a par with each other (Srinivasan 1989) on a unit-time basis.

The development of the short-duration variety is relatively more recent. A short-duration variety can be defined (Sahi and Sundara 1986) as one which has the following characteristics:

1. It matures in 8 months and records 16% pol in juice and 85% purity when planted in January to February.
2. It produces at least 85% of the average cane yield of the ruling 12-month variety.

Table 2.4. Yield levels of early and mid-late varities. (Srinivasan 1989)

Variety	Yield ($Mg\,ha^{-1}$)	Variety	Yield ($Mg\,ha^{-1}$)
Early group (10 months)		Mid-late group (12 months)	
CO775	97.6	CO419	111.7
CO997	67.5	CO62175	99.2
CO62174	89.0	CO62198	126.8
CO6806	98.2	CO6304	124.6
CO7704	64.3	CO7717	111.8
COC671	85.0	CO1148	96.0
COJ64	53.4		
Mean	79.3		96.0
Production per month	7.93		8.0

3. It maintains the quality for 1–2 months.

During 1983, six short-duration varieties, namely CO8336, CO8337, CO8338, CO8339, CO8340 and CO8341, were released for field testing in peninsular India. The most promising was CO8341 (CO7507 × CP34-79) which contained 18.58, 89.9 and 11.96% sucrose, purity and fibre, respectively, after 240 days. Variety CO8338 was released in the state of Gujarat to replace COC671 in wilt-affected soils as the latter is highly susceptible to wilt disease.

We look forward with interest to the development of early/short duration rich canes which will help extend the grinding period, improve overall sugar recoveries, reduce cultivation costs and, above all, fit into the multiple cropping system.

2.7 Varietal Decline

Abbot (1959) defined the yield decline as the 'gradual loss of vigour and yielding ability over a period of years of commercial cultivation, the causes of which are often obscure'. This is also termed the 'run-out' of varieties. After a few years of cultivation the commercial variety fails to maintain its original yield level. The yield decline is often associated with biotic and abiotic stresses. Humbert (1959) demonstrated that a lack of aeration, reduced O_2 supply and restricted root proliferation following soil compaction are the causes of varietal decline. Similarly, soil erosion, soil compaction by heavy equipment, depletion of trace elements, a lack of organic matter, poor drainage causing anaerobic conditions and an accumulation of toxic elements in the soil might cause the yield decline. Pests and diseases could also be major factors of yield decline in cane. Ratoon-stunting disease (RSD), mosaic, nematodes, red rot, chlorotic streak and *Pythium* root rot have been responsible for yield decline in many important varieties and thus had to be replaced by better varieties for commercial cultivation. Viruses have also been implicated in the 'run-out' of a few varieties and the chief vector/carrier is the ubiquitous pink mealy bug (*Saccharicoccus sacchari*). Variety CO419/CO740 was under cultivation for more than four decades in India but its use was faded out due to yield decline. The decline is faster when the variety is grown in sandy soils without organic matter and is subject to poor management. The visible symptoms of yield decline are that the lower leaves become chlorotic and senesce very early (6–8 months) and the crown has five to six small green leaves with a tapering apex.

Yield decline seems to be a 'syndrome' and many factors are responsible for the gradual reduction in yield and quality. However, a suggestion is mooted that heat therapy to setts may rejuvenate the variety. But the ultimate solution lies in the release of superior varieties for general cultivation.

2.8 Some Applications of Biotechnology to Sugarcane[2]

It is said that man started domesticating plants some 10 000 years ago. He has gone a long way on the sands of time from hunting, gathering and shifting cultivation to intensive crop production and crop simulation models. However, during the past 20 years, the impact of biotechnology on agriculture has been tremendous. Exciting possibilities for crop improvement, including sugarcane, are available and a quantum leap in crop production is possible through prudent application of biotechnology. However, biotechnology does not offer 'zero risk', thus ethical issues and environmental risks are genuine and must be addressed.

Biotechnology is a general term applied to any of a broad range of disciplines by which living systems are modified or manipulated to produce novel products, enhance plant and animal productivity, mediate disease resistance, etc. Baker (1991) naively defines biotechnology as the application of living organisms to create products and processes. It represents the most exciting advance in biological sciences of this century. According to Baker, the word 'advance' is important for it implies a continuum. In other words, biotechnology is itself not new. The production of alcohol or yeast from molasses and the biological delignification of bagasse represent applications of biotechnology. Thus, biotechnology has increased our ability to reduce costs and to enhance the efficiency of production. Moreover, developing countries with small and marginal farms stand to gain from such advances.

In sugarcane the pioneering work has been done by Heinz and Mee (1969), who first demonstrated that plantlets could be developed from cane callus[3] culture. Hendre et al. (1983) cultured shoot tips of cane cultivar CO740 and produced about 200 000 plants in 6 months from a single shoot tip. These plants have been field tested. They observed that clonally propagated plantlets were similar to the conventionally sett-propagated plants with regard to millable cane length, number of internodes and number of canes per clump. They reiterated that micropropagation is an important means to produce a larger number of elite clones by *in vitro* culture. Furthermore, plants regenerated from callus may display a wide range of characteristics. Such variation is called somoclonal variation. This variability is used to select lines of desired traits, e.g. high sugar, cane yield, disease resistance, etc. In Taiwan somaclones of sugarcane have been selected for some of these traits.

Anther culture and haploid breeding have proved to be a powerful tool in developing new varieties. These haploid sporophytes can be used for mutation. By chromosome doubling, a large number of homozygous lines (dihaploids) can be produced in a short time. In China anther-derived haploids of sugarcane have been obtained.

---

[2] Detailed treatment of this subject is beyond the scope of this book.

[3] Callus, plant cells which proliferate (grow and divide) on a defined medium and lack morphological differentiation.

Wide hybridization involving intergeneric crosses have been attempted to improve sugarcane. Some crosses have been successful in the past between *Saccharum* sp. × *Zea* sp., *S. officinarum* × *Sclerostachya fusca* and *Saccharum* sp. × *Ripidium arundinaceum*. However, these crosses invite sexual incompatibility, sterility of hybrids and other barriers that frequently occur. To overcome such barriers two alternatives are employed, namely embryo rescue and somatic hybridization (Sinha 1992). In the former, embryos are rescued by growing them on artificial tissue culture. In sugarcane the technique appears to be labour-intensive (Sinha 1992). The somatic hybridization is accomplished by fusing protoplasts of two donor parents. Protoplasts are obtained by the enzymatic (cellulase and pectinase) digestion of the cell wall. The intergeneric protoplast fusions are useful for introducing genes from distantly related genera which are amenable to culture; thus, new varieties with desirable traits are produced.

Tissue culture, an important branch of biotechnology, involves the growth of the plant or animal cell tissues and organs outside of the organism, usually in glass or plastic containers (in vitro culture). Here, the 'explant' tissue or cells are supplied with nutrients and other substances required for continued growth. In vitro selections offer inexhaustible possibilities to screen clones for resistance to biotic and abiotic stresses. In sugarcane great strides have been made by this technique in selecting some disease-resistant varieties.

Eye spot disease produces the toxin helminthosporoside. This purified toxin was incorporated in tissue culture medium and the surviving callus cells were selected to produce eye spot-resistant clones. Efforts have also been made to select resistant clones for a very serious disease, i.e. red rot, which so far has not been controlled by conventional breeding methods. Similarly, salt-tolerant lines in sugarcane have been identified. In Thailand virus-free planting material of sugarcane has been developed. Also, elite clones can be selected to resist adverse soil conditions such as iron toxicity, zinc deficiency or acid sulphate. Another serious disease afflicting sugarcane, called grassy shoot (GSD), is caused by mycoplasma-like organisms (MLOs), synonymous with spiroplasma. The spiroplasmas are prokaryotes without a cell wall; they are helical and mobile and are important pathogens of plants. Monoclonal antibodies (McAbs) have been generated which are highly specific to spiroplasma and help as probes (diagnostic purposes) in detection and identification.

Relatively recent research shows that associative N fixation (rhizocoenosis) is assuming greater importance and thus helping to reduce costly input, e.g. fertilizer N. It has been observed that *Azospirillum brasilense* is present in the root zone of sugarcane. These free N-fixing organisms can be grown in the callus cultures of sugarcane. Selection of plantlets from callus culture with a high rate of associative N fixation helps greatly in reducing the N dosage, which is a key factor for yield and quality. It is apparent that the introduced strains of *Azospirillum* are better than the wild ones. Further, the population of mutant *Azospirillum* may have a greater rhizo-plane (soil-free root surface) and histosphere (within the root zone).

Presently, recombinant DNA[4] technology is available to alter the plants. Transformation of plants by *Agrobacterium tumefaciens*-mediated DNA transfer is most commonly used to accomplish plant gene transfer. In *Agrobacterium* transformation, genetic sequences are introduced into modified Ti-plasmids (tumor-induced plasmids) which carry additional genetic elements required for the DNA-transfer process. Unfortunately, gene transfer from this bacterium is not possible in graminaceous plants, including sugarcane. Hence, Weissinger (1992) suggests physical methods of gene transfer in sugarcane which include inter alia, microinjection[5], liposome fusion[6], electroporation[7] and microprojectile bombardment or biolistic gun DNA transfer.[8] Of these, electroporation and biolistic gun transfer are being used in Australia to produce transgenic sugarcane plants.[9] A synthetic gene conferring resistance to sugarcane mosaic virus has been introduced in sugarcane cells by electroporation.

One of the most novel applications of biotechnology is to induce resistance of plants to herbicides. There are several methods by which cells become resistant to herbicides:

1. inhibition of uptake;
2. degradation;
3. conjugation;
4. sequestration;
5. overproduction of competing substrates or metabolites;
6. overproduction of target metabolites; and
7. alteration of the target protein.

From these methods, two strategies are especially important to obtain herbicide-resistant plants:

1. decreasing the sensitivity of the plant to herbicides by the overproduction of target enzyme or by expression of a modified target that is insensitive to the herbicide but retains its enzymatic activity;
2. providing the plant with a herbicide-detoxifying pathway.

---

[4] Recombinant DNA, the hybrid DNA formed by combining pieces of DNA from different organisms, e.g. a bacterial plasmid containing a plant DNA sequence, is a recombinant DNA molecule.

[5] Microinjection, DNA delivery technology in which DNA is injected into cells or protoplasts with a microscopic needle drawn from a fine glass tube.

[6] Liposome fusion, liposomes, synthetic single or multi-lamellar lipid vesicles, have been used successfully as 'vehicles' for both RNA and plasmid DNA into protoplasts.

[7] Electroporation, the introduction of transforming DNA or RNA into protoplasts or other cells by the momentary disruption of the cell membrane through exposure to an intense electric field.

[8] Microprojectile bombardment or biolistic gun DNA transfer, this is a process in which transforming DNA is associated with microscopic metal (tungsten or gold) particles that are accelerated to high velocity in a particle acceleration apparatus (particle gun).

[9] Transgenic plants, stably transformed plants with cloned genes of interest to agriculture and industry.

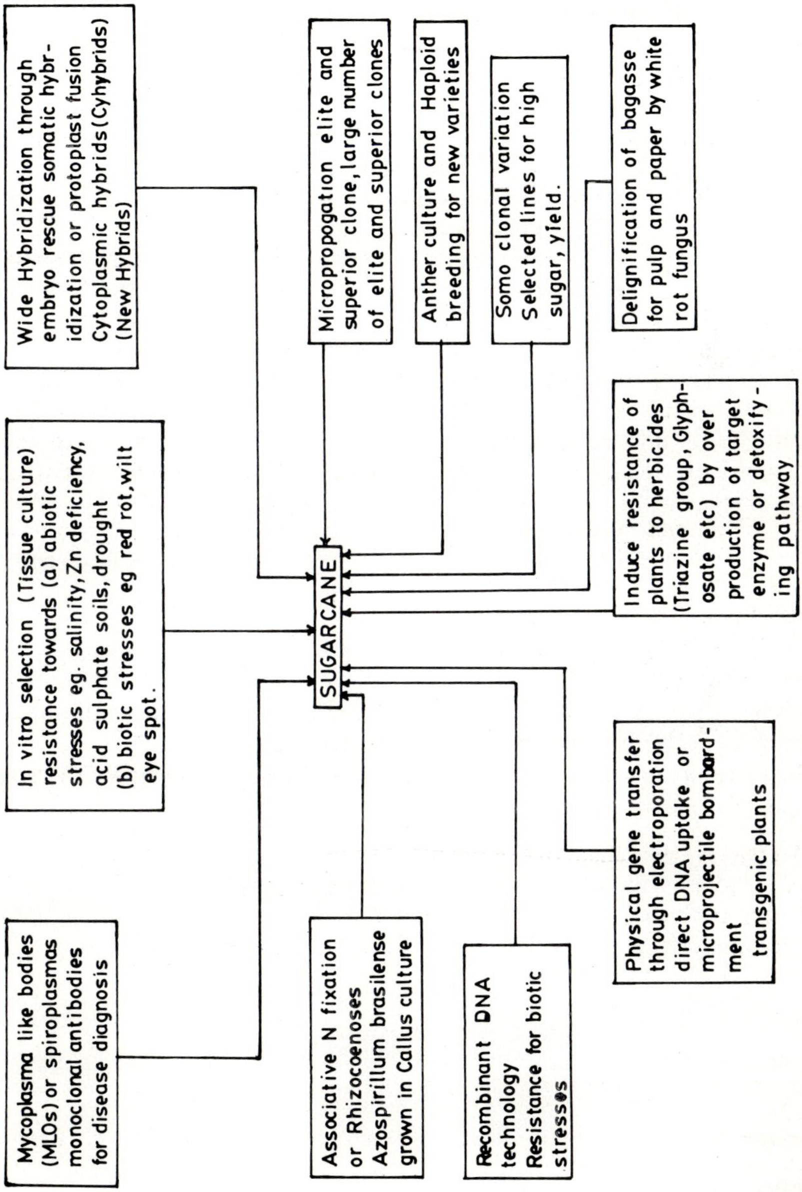

Fig. 2.6. Some services offered by biotechnology to sugarcane

Examples of the first approach are the engineered, resistant plants to glyphosate and sulfonyl urea compounds. Glyphosate is an inhibitor of 5-enolpyruvylshikimate-3-phosphate synthase (EPSPS), an enzyme in the biosynthetic pathway of aromatic amino acids. Sulfonyl urea herbicides inhibit acetolactate synthase (ALS), an enzyme involved in the biosynthesis of branched-chain amino acids. Thus, overproduction of EPSPS and ALS confer to transgenic plants[9] tolerance to glyphosate and sulfonyl urea respectively.

Another broad-spectrum herbicide glufosinate or 1-phosphinothricin (PPT) is used in many crops including sugarcane. A PPT-resistant gene 'bar' has been isolated which converts PPT into an acetylated, non-toxic form; this helps the plant to take the detoxifying pathway. Besides the above-mentioned herbicides, the triazine group (e.g. atrazine) is also extensively used in sugarcane to control weeds. Plants can tolerate atrazine because they are able to conjugate the herbicide with glutathione (a protein involved in photosynthetic electron transport) before the herbicide reaches the site of action. Thus, with transgenic plants, post-emergence application of potent, but environmentally safe herbicides can be used in sugarcane.

A direct application of biotechnology to the sugar industry is discerned from selective delignification of bagasse for pulp and the production of paper. The white rot fungus (*Phanerochaete chrysosporium* = *Sporotrichum pulverulentum*) of Basidiomycetes is an efficient delignifier. However, it is feared that the use of lignocellulose for the production of ethanol or chemical feedstock will prove to be a difficult task and a dramatic break-through by recombinant DNA belongs to the distant future.

Some important applications of biotechnology to sugarcane are shown in Fig. 2.6.

Chapter 3 Soil and Climatic Requirements

3.1 Soils

Sugarcane is not that exacting with regard to its soil requirements. It may be grown in a variety of soils with textures ranging from sand or heavy clay soils to organic soils. Tabayoyong (1959) found no difference in the yield/sugar percentage when sugarcane was grown in clay, clay loam and sandy clay-loam soils in the Philippines. In West Africa cane is grown in the seasonally flooded wetlands along river banks called *fadamas*. Ideal sugarcane soils are deep, aerated, well-structured sandy loams to clay loams with an adequate supply of organic matter.

3.1.1 Soil Classification

A soil is conceived as consisting of structural units called 'pedons' (pedon = ground) with contiguous similar pedons termed polypedons (Soil Survey Staff 1975). Figure 3.1 depicts the predominant sugarcane soils of the world and these are briefly described in Table 3.1.

Sanchez (1976) has indicated that Oxisols (highly weathered), Aridisols (desert soils) and Alfisols (red soils) constitute nearly 60% of the area in the tropics where sugarcane is concentrated. Oxisols or Latosols are the predominant soils of humid and subhumid regions and have rainfall from 250 to 10000 mm per annum. The predominant clay is kaolinite with a cation exchange capacity (CEC) of 7–12 meq. 100 g^{-1} soil.[1] Large areas of Oxisols occur in South America, East–Central Africa, Cuba, parts of India, Sri Lanka, Thailand, Vietnam, etc. Two soil orders, namely Histisols and Andisols, deserve special attention.

Histisols are organic soils (bog/half bog soils) that are dark brown/black, have a low bulk density, high water-holding capacity and a porous, open structure. Indonesia has 12 million ha of organic soils while Florida and Louisiana have 2 million ha (Brady 1984). Barnes (1974) has narrated the development of the peat soils of Florida for cane cultivation. A standard dressing of 75 kg $MnSO_4$ and 28 kg each of $CuSO_4$ and $ZnSO_4$ per hectare is applied at the time of planting. In addition, new varieties have improved the yield and quality of cane.

Andisols (Andepts) are not extensive in the tropics but are found with deposits of volcanic ash. They are extensive in the Caribbean, Columbia,

[1] meq 100 g^{-1} soil = 1 centimole [C mol] kg^{-1} soil.

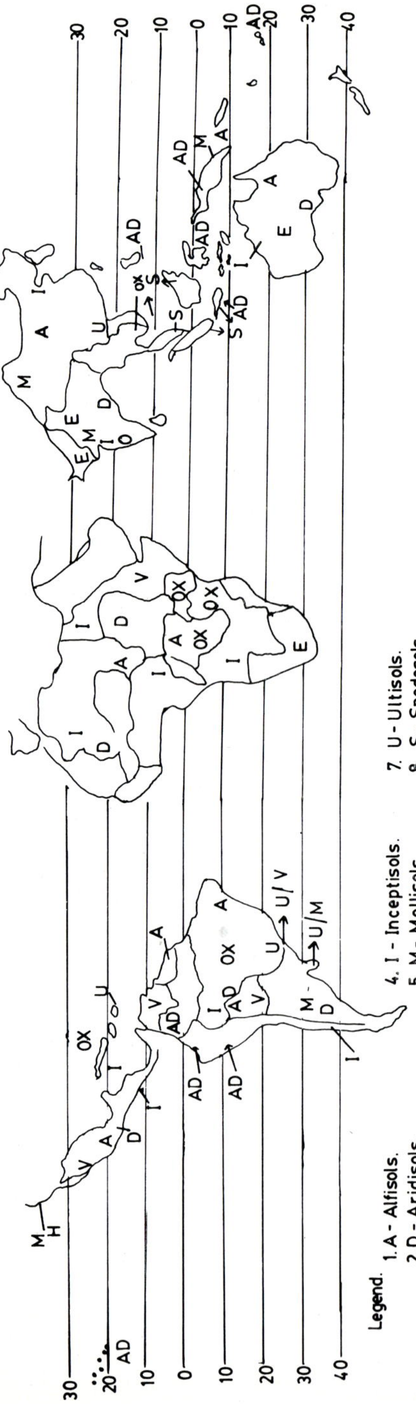

Fig. 3.1. Soil orders of some important sugarcane-growing countries in the world

Table 3.1. Brief description of the main soil orders with some important sugarcane areas of the world

Soil order	Brief description	Some important sugarcane areas
Oxisols	Rainfall 250–10 000 mm per annum, highly weathered, predominant clay = kaolinite with CEC 7–12 mEq. $100\ g^{-1}$ soil	South America, East–Central Africa, Cuba, parts of India, Sri Lanka, Thailand, Vietnam
Aridisols	Desert soils, PET[a] far exceeds precipitation, soil moisture most limiting. Cane grown under irrigation	South central Australia, Mexico, southern Argentina, West Africa
Alfisols	Red soils, low base saturation, responsive to management	East-Central Africa, Central China, Eastern Brazil, Southeast Asia, parts of Australia
Ultisols	Argillic (clay) horizon, base saturation 20%, deep well-drained red and yellow soils, poor physical properties, low native fertility, dominant clays 1:1 type, oxides of Fe and Al, responds to management	Coastal Australia, Puerto Rico, Southeast Asia, southern China, Brazil, Paraguay
Inceptisols	Moderately developed soils of humid regions, soil horizons with no extreme weathering. Two suborders; Tropepts are tropical Inceptisols and Aquepts are wet Inceptisols	Chile, North Africa, eastern China, northwestern Australia, Central Africa, southwestern India, Brazil. Aquepts found along Amazon and Ganges rivers
Entisols	Weakly developed mineral soils	South Africa, central Australia, North India, Pakistan
Vertisols	'Cracking'. Clay soils only 2% of tropical area; sticky and plastic when wet, hard when dry	Venezuela, Bolivia, Paraguay, Mexico, Ethiopia, parts of Australia, Africa, India
Mollisols	Base-rich, organic, not peaty surface horizon (mollic epipedon), only 1% of tropical soils	Argentina, parts of China, Mexico, Paraguay, Uruguay, North India
Histisols	Bogs/half-bog soils, dark brown/black colour, high CEC and water-holding capacity	Southeast Asia, Florida, Louisiana, Papua New Guinea
Spodosols	Podzols, minor importance, highly leached, acidic, coarse-textured, infertile, low in nutrient status, CEC and water-holding capacity	Indonesia, Malay Peninsula, Sri Lanka
Andisols	Andepts, volcanic ash soils, dominant mineral, allophane	Columbia, Ecuador, Peru, Bolivia, Indonesia, Papua New Guinea, Hawaii

[a] PET = potential evapotranspiration.

Ecuador, Peru, Bolivia, Indonesia, Papua New Guinea and Hawaii. The dominant mineral is an allophane – an amorphous aluminium silicate which forms complexes with organic matter (Andepts). These soils are free draining, low in bulk density and have an excellent stable structure and high infiltration rates.

3.1.2 Physicochemical Properties of Soils Influencing Sugarcane Growth

The physical properties of soils which greatly affect cane growth are: structure, bulk density, soil atmosphere, thermal properties and water retention and transport characteristics.

3.1.2.1 Structure

Soil is an aggregation under natural conditions of primary particles, like sand, silt and clay, into larger units called peds. This is in contrast to clods which are caused by disturbances in ploughing or digging. For better yield potential, sugarcane requires a crumb soil structure but due to mechanized cultivation plate structures are often encountered and these are broken by deep cultivation.

3.1.2.2 Bulk Density

The effect of bulk density (BD) on sugarcane roots is well documented and the critical value seems to be 1.8–1.9 g cm^{-3}. Trouse and Humbert (1961) demonstrated that in Hawaiian soils as BD increased, roots became flattened and restricted. An interaction between BD and soil type was discernable. In Andisols (Andepts) a BD of 1.08 g cm^{-3} restricted root growth, compared with 1.52 g cm^{-3} in Latosols. Furthermore, as BD increased so root efficiency decreased, as evidenced from the reduced uptake of ^{89}Rb (Trouse and Humbert 1961).

A detailed review by Vomacil (1957) indicated that a high BD has a significant influence on mechanical impedance and root growth with a consequent reduction in water availability and root aeration. Ricaud (1977) showed that there was a severe reduction in cane yield in the sandy soils of Louisiana due to soil compaction. This is confirmed by the findings of Monteith and Banath (1965) who stated that as BD increased, mechanical impedance became dominant. Thus, soil strength, as measured by penetrometer, controlled root growth. Hence, penetrometer readings prove to be a useful diagnostic tool for indicating root proliferation and extension (Table 3.2). The suggestion has been mooted that ethylene production mediates the plant response to soil strength and endogenous ethylene increases as BD increases (Jones 1985).

Table 3.2. Effect of soil compaction and soil strength on the percentage of macropores and the root growth of sugarcane (*S. officinarum* cv. Triton) grown in different soil types. (Monteith and Banath 1965)

Soil	Bulk density ($g\ cm^{-3}$)	Penetrometer reading ($kg\ cm^{-2}$)	Macropores (%)	Root dry matter (mg)	Root:Shoot ratio
	1.73	4.50	0	0	0
	1.50	1.73	0.46	76	0.58
Loam	1.35	0.69	1.91	115	1.84
	1.15	0.23	5.98	125	1.36
	1.74	4.50	0	0	0
Clay loam	1.52	2.19	0.52	51	0.55
	1.33	0.75	1.82	131	1.40
	1.17	0.28	5.73	91	2.10
	1.46	1.08	0	10	0.14
Silty clay	1.38	0.87	0.56	33	0.57
	1.27	0.62	5.22	65	1.17
	1.17	0.34	7.67	96	1.64

Mechanical strength reduced the foliar concentration of nutrients and their accumulation in the plant (Trouse and Humbert 1961). In the studies of Juang and Uehara (1971) soil compaction reduced P and K uptake by cane as well as reducing the dry weight of roots (Table 3.3). As BD increased the reduction in P uptake (90%) was found to be more severe than the reduction in K uptake (50%).

This author is of the opinion that for most sugarcane soils, a BD of 1.5 $g\ cm^{-3}$ in the surface soil and a BD of 1.6 $g\ cm^{-3}$ in the subsurface soil forms a critical limit for good yield and juice quality.

Table 3.3. Effect of bulk density on phosphorus (^{32}P) and potassium (^{86}Rb)[a] uptake and sugarcane root growth in a clay loam soil. (Juang and Uehara 1971)

Bulk density ($g\ cm^{-3}$)	Phosphorus		Potassium	
	Root dry weight in a core (g)	^{32}P uptake ($cpm\ g^{-1}$ soil)	Root dry weight (g)	^{86}Rb uptake ($cpm\ g^{-1}$ soil)
1.2	0.46	385	0.42	10.0
1.4	0.26	274	0.36	12.0
1.6	0.18	73	0.30	4.5
1.8	0.15	35	0.05	5.5

[a] Half-life of radioactive K is only a few hours; ^{86}Rb is used in place of K.
cpm = counts per min.

3.1.2.3 Soil Atmosphere

Soil pores which are devoid of water and are filled with gas constitute the soil atmosphere. The soil atmosphere is poorer in O_2 and richer in CO_2 than the free atmosphere due to microbial activity and root respiration; it is also influenced by soil water and temperature. Gaseous transfer takes place between the free atmosphere and the soil atmosphere through diffusion and the relationship is given as:

$$D/Do = \alpha S \approx 0.6\ S,$$

where D is the diffusion coefficient of gas in soil, Do is the diffusion coefficient of gas in air and S is the proportion of the soil volume occupied by air. For most plants an oxygen diffusion rate (ODR) of 3–4 mg $O_2\ cm^{-2}\ s^{-1}$ is critical (Payne and Gregory 1988).

Sugarcane is sensitive to an O_2 deficiency for this seriously impairs root development and nutrient uptake. Banath and Monteith (1966) observed that an O_2 level of 3.4% at the soil surface was critical and the sharpest decline in root weight was noted between 3 and 0% O_2 (Table 3.4). An interaction between soil BD and the O_2 concentration was evident. As BD increased, a higher concentration of O_2 was required to maintain normal root growth.

Jones (1985) concluded that root responses to soil strength and poor aeration are similar and that ethylene production mediates such responses.

3.1.2.4 Thermal Properties

Soil temperature is seldom a serious limiting factor in the tropics and it approximates the air temperature to a depth of 50 cm (Sanchez 1976). Of the thermal properties, thermal conductivity (K) and thermal diffusivity are important. Thermal conductivity (K) is 'that property of soil which relates to the transfer of heat away from or towards soil surface' (Richards et al. 1952). It is dependent on the air and water content of the soil. A compact soil with a higher water content tends to have larger K values.

Table 3.4. Effect of oxygen supplied to the soil surface on the growth of sugarcane roots. (Banath and Monteith 1966)

O_2 supplied to soil surface (%)	Root dry weight (g)
0.0	0.13
2.1	0.13
3.4	0.34
7.8	0.52
14.6	0.60
Air	0.54

Thermal diffusivity denotes the temperature change that takes place in any portion of soil as the heat flows into it from an adjacent layer (Stolzy and Jury 1982).

Soil temperature less than 10 °C reduces the water uptake even when the soil moisture is sufficient. This is called physiological drought. In many sugarcane areas the soil temperature ranges from 16 to 32°C. However, in the cooler winters of Iran, Louisiana, Pakistan, subtropical North India, etc. cane growth is retarded. But mulching with polyethylene plastic in the Oxisols of Hawaii and Aridisols of North India has improved the yield of cane substantially. Through the 'greenhouse' effect these mulches have increased the mean soil temperature and reduced the irrigation requirement. In sugarcane areas where soil temperatures are high, soil cover with stubble or trash mulch will lower the temperature, increase soil moisture storage and decrease competition from weeds. Thus, control of soil temperature should form an important component of cane management.

3.1.2.5 Water Retention and Transport Characteristics

The size and shape, or geometry, of soil pores controls to a large extent the retention, availability and rate of water movement. The pore-size distribution in soil depends on the structure and stability of aggregates. Water moves freely through pores wider than 0.3 mm and the young roots can move easily through these pores and absorb water. Different pore sizes and the suction required to empty them of water is given in Table 3.5

Quoting D. J. Greenland, Payne (1988) classified water in pore sizes of 50, 5 and 0.5 μm as being transmission, storage and residual respectively. As sugarcane is a deep rooted crop of long duration, all three water components are important for good growth and yield. For most soils, the field capacity is taken as 0.03 MPa (0.33 bar) and the wilting point as 1.5 MPa (15 bar) and the difference in water content between these two soil constants is called the available water. Sugarcane draws water easily between suction values of 20 and 250 kPa (0.2–2.5 bar).

Besides water retention and transport, the infiltration rate is equally important. In general, Oxisols, Ultisols and Mollisols have high infiltration rates ranging from about 7.5 to 8.5 cm h^{-1}; this reflects their good structure.

Table 3.5. Suction required to empty soil pores of water. (Payne 1988)

Pore size (μm)	Suction required to remove water (kPa)
300	1(0.01) [a]
30–60	5–10 (0.05–0.1)
1	300 (3)

[a] Values in ( ) denote suction in bar.

By contrast, Vertisols have low values of 0.1–0.5 cm h^{-1} due to high clay contents dominated by expansible 2:1-type clays.

3.1.3 The Rhizosphere

It is increasingly recognized that non-leguminous plants can also fix nitrogen and such associative N fixation is termed 'rhizocoenoses'. The aerobic diazotroph *Azotobacter paspalum* in association with the roots of *Paspalum notatum* cv. batatais, a tropical grass, can fix N to the extent of 90 kg ha^{-1} yr^{-1} and this is comparable to *Rhizobium*–legume symbiotic N fixation. In general, the associative N fixation by grasses and some crops ranges from 5 to 20 kg $ha^{-1}yr^{-1}$ but sugarcane and rice are notable exceptions which receive higher quantities of microbially fixed N (Zuberer 1990).

The term rhizosphere (Greek *rhizo* = root, sphere = natural surroundings) was first introduced in 1904 by Hiltner who defined it as the interfacial volume between plant roots and the bulk soil in which the microbial population lives. The rhizosphere is where root meets soil. Lynch (1990) described, in a comprehensive treatise, the total rhizosphere environment by an interacting trinity of soil, plant and organism. The boundaries of the rhizosphere are not exact (Fig. 3.2.). Some fungal hyphae live in the soil and penetrate the outer cortical cells of roots (endorhizosphere) while others colonize on the root surface (ectorhizosphere).

The rhizosphere may be up to 1–2 mm thick and the concept of a rhizoplane has been envisioned (Fig. 3.2). The microbial population surrounding the root surface is 200 times larger than in the bulk of the soil, but the root tip is typically devoid of microorganisms.

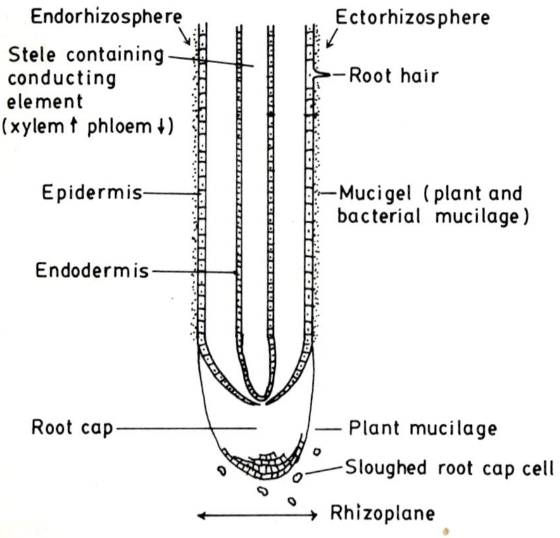

Fig. 3.2. The rhizosphere. (Lynch 1990)

The rhizosphere affects plant growth by (1) rhizodeposition (sugars, amino acids, growth regulators, vitamins, hormones, etc.), (2) production of pseudobactin, known as plant growth promoting rhizobacteria (PGPR), (3) formation of stable soil aggregates, and (4) production of siderophores which increase P availability and enhance Fe availability through chelation (Lynch 1990). The poor rot syndrome (PRS) of sugarcane, caused by *Pythium graminicola*, may possibly be controlled by the rhizosphere bacteria (Hornby 1990). Increased resistance of cultivar CO453 to *Pythium* root rot is attributed to the antibiotic properties of the rhizosphere (Srinivasan 1968).

The availability of immobile P and other macro- and micro-nutrients is increased by the symbiotic associations, termed mycorrhizas, between plants and specialized fungi. These are classed as ecto-or endomycorrhizas depending on the physical relationship between the fungus and plant root. Of great interest are vesicular-arbuscular mycorrhizas (VAM) which are known to improve soil–plant relationships, especially those of C_4 plants where it leads to increased disease resistance of the plants and improved nutrient availability (Reid 1990).

It has been suggested that the rhizosphere of C_4 plants is colonized by *Azospirillum lipoferum* and that of C_3 plants by *A. brasilense*. But Dobereiner (1961) conclusively demonstrated that *Beijerinckia* colonized the sugarcane rhizosphere and enhanced N fixation. A high sucrose exudate from the cane was responsible for increased activity of *Beijerinckia* with a concomitant reduction in the number of bacteria, actinomycetes and fungi in the sugarcane rhizosphere. But in acid soils (pH 3) *Saccharobacter nitrocaptans* was associated with the rhizosphere of sugarcane (Cavalecante and Dobereiner 1988).

The difference in N fixation by the rhizosphere of different sugarcane varieties has been alluded to by Ruschel and Ruschel (1977). They observed that NA56-79 (683.6 nmol ethylene evolved/plant) and CB46-47 (551.1 nmol ethylene evolved/plant) had significantly higher N fixation than other varieties. The high sucrose content of these varieties is associated with higher microbial colonization, confirming the trends shown by earlier investigations. A popular commercial variety, CB41-76, was low in sugar and had the lowest N fixation (84.9 nmol ethylene evolved/plant).

Throughout the world sugarcane is well manured and fertilized and the activities of rhizospheric diazotrophs give new possibilities for reducing chemical inputs, which have a negative impact on the environment, while maintaining sustainable productivity of both cane and sugar. It is even conceptualized that suprarhizosphere organisms could be developed which would greatly modify the soil conditions and lead to better plant growth (Lynch 1990).

3.1.4 Problem Soils

Abiotic stresses such as saline, sodic and acid soils impose great restrictions on growth and cane and sugar yields and therefore special management skills are required. The optimum pH for sugarcane growth and development is 6–8.

3.1.4.1 Saline–Sodic Soils

These soils are more frequently found in arid and semi-arid tropics where sugarcane is extensively grown. Globally, about 1 million ha are affected by salinity and/or sodicity. In these regions evapotranspiration far exceeds precipitation and salts accumulate in the soil horizon. The main sources of salts are ‘fossil salts’ (natural soluble salts of soil are NaCl, $CaCl_2$, $MgCl_2$, $CaSO_4$, $MgSO_4$, etc.) and various surface and groundwaters. The most pronounced effect of saline soil is the increase in total water potential (ψ) with a consequent reduction in water and nutrient uptake:

$$\psi_0 = \psi_g + \psi_c + \psi_o$$

(Total water potential) = (gravitational potential) + (capillary potential) + (osmotic potential).

The US Salinity Laboratory Staff (1954) has given the following relation:

$$\psi_0 \approx 0.036\ \mathrm{EC},$$

where ψ_0 is the osmotic potential (MPa) and EC is electrical conductivity (d S m^{-1}). (Note that d S m^{-1} = m S cm^{-1} at 25 °C). Saline soil contains excessive quantities of soluble salts (EC= > 4d S m^{-1}, pH < 8.5) as compared with excessive adsorbed sodium in sodic soil (exchangeable sodium percentage, ESP= >15, pH > 8.5). Sugarcane is moderately tolerant to salts and can tolerate an EC up to 1.7 d S m^{-1}; this is in contrast to rice and barley which can tolerate EC values up to 3 and 8 d S m^{-1}, respectively. An excessive salt concentration results in poor germination, an uneven stand, stunted growth, chlorotic leaves and development of poor stools which can be easily pulled with a dead root mass (Barnes 1974). Saline–sodic soils impair the juice quality and result in reduced sugar recovery, higher reducing sugars and higher yields of molasses. Under these stress conditions, a deficiency of Fe, Mn, Cu and Zn is more frequently experienced. Even tassling is induced when the salinity increases the osmotic pressure to 200 kPa (2 bar).

For optimum cane and sugar yields, the salinity level should be kept below 2 d S m^{-1}, and the threshold EC value seems to be 4 d S m^{-1} (Mehrad 1968). Bernstein et al. (1966) described that yield decrements of 10 and 25% were attained at EC values of 3 and 5 d S m^{-1}, respectively. Growth practically ceased at 8.6 d S m^{-1} conductivity. In Iran, Shoji and Sund (1965) obtained cane yields of 101.9 and 48.2 Mg ha^{-1} at EC values of < 2 and 5.2 d S m^{-1}, respectively. The interaction between salinity levels and edaphic factors on yield and quality of cane was stressed by Fogliata and Aso (1965). The authors recognized the specific ion effects of Cl^-; SO_4^{2-} and CO_3^{2-} which, when combined, reduced the yield by up to 80%. The effect of soil salinity on cane yield was succinctly documented by Nour et al. (1989) under Egyptian conditions (Table 3.6). Fine-textured soils were associated with a high EC and an increase of one unit of EC (1 d S m^{-1}) resulted in a decrement of 5.45 Mg ha^{-1} of cane ($r = -0.88$). The degree of salinity was highly pronounced in clay soils with imperfect drainage.

The crop stage is equally important, with germination and tillering being more sensitive to salinity than the later stages.

Varietal differences in salt tolerance have been recognized. In general, high-fibred varieties were more tolerant; NCO310 was less tolerant than CP44-101 (Mehrad 1968) and NCO293 (Bernstein et al. 1966). B42231 in Jamaica and TUC2645 in Argentina were more tolerant to saline conditions. Other varieties which were tolerant to salt-shock tests were H50-775, H50-7209, H57-5174 and CO6806.

Field and greenhouse tests suggest that sugarcane is not affected at EC less than 2 d S m^{-1}, and an EC of 4 d S m^{-1} is a threshold value beyond which yield and quality are impaired. Processing problems of cane grown under saline conditions are associated with higher K contents.

Summarizing the results of several investigators, Blackburn (1984) conjectured that the limits of salinity on cane growth and yield are as shown in Table 3.7.

Short-term measures to reclaim saline-alkali soils include scraping off the surface soil, removal of lime concretions if present, organic matter addition

Table 3.6. Effect of soil types and salinity levels on cane yield. (Nour et al. 1989)

Salinity range (d S m^{-1})	Mean salinity (d S m^{-1})	Soil category	Soil texture	Yield (Mg ha^{-1})	Yield (%)
0.99– 2.60	1.62	Non-saline Non-alkaline Normal	Clay and loam	93.8	100
4.96–10.46	6.98	Saline	Clay and loam	51.3	54.7
13.88–17.32	15.6	Saline-alkaline	Clay	21.7	23.2

Table 3.7. Effects of salinity on cane growth

	Cane healthy	Cane slightly affected	Cane severely affected	Cane dead
SAR	13	18	35	41
pH	8.2	8.3	9.1	9.2

SAR, sodium adsorption ratio = $Na/(Ca + Mg)^{1/2}$.

and green manuring. Agronomic manipulation lies in planting cane in the furrow or on the side of the furrow so that salts accumulate on the top of the ridge without affecting cane growth. Long-term measures include reclaiming saline soils by copious irrigation and reclaiming sodic soils by addition of gypsum/sulphur with copious irrigation to remove Na^+ from the exchange complex. However, the sure solution rests with the development of saline–sodic tolerant varieties.

3.1.4.2 Acid Soils

Sanchez (1976) suggests that tropical America has more acid soils than tropical Asia/Africa. Soil acidity is the sequel to the presence of Al^{3+} in soluble and exchangeable forms. Leaching of bases takes place with the dominance of exchangeable Al^{3+} and a small amount of exchangeable hydrogen. Rowell (1988) proposes that a large proton input (H^+) will replace cations like Ca^{2+}, Mg^{2+}, K^+, etc. with the resulting development of an acid soil. It is contended that this soil reaction per se is not the sole factor but aluminium toxicity is responsible for reduced yield and cane quality.

The adverse effects of acid soils on sugarcane have been well described (Evans 1965). Aluminium toxicity is characteristic of pegasse soils (tropical peat) in Hawaii, Guyana and Trinidad and of Ultisols in Puerto Rico and Brazil. The most visible injury caused by aluminium in cane is a stubby, restricted, coralloid root system. Acid soils cause a severe deficiency of Ca, Mg and P and lead to toxic levels of Mn and Fe in cane tissues. In Hawaiian soils with little exchangeable Al^{3+} (below pH 5) liming is used more as a Ca^{2+} fertilizer than as an amendment.

In the tropical Ultisols of Puerto Rico (Abruna-Rodriguez and Vicente Chandler 1967) sugarcane (cv. PR980) yield was drastically reduced due to low base saturation and exchangeable bases and higher exchangeable Al^{3+} (Table 3.8). However, yields improved fourfold following liming.

Commonly used liming materials are limestone, dolomite limestone, burnt lime and slaked lime. The efficiency of these materials depends on their fineness and a good grade is 60–100 mesh.

Morelli et al. (1989) have recently shown that a combination of lime and phosphogypsum increased the base saturation and reduced exchangeable

Table 3.8. Cane yield in the acid soils of Puerto Rico. (From Abruna-Rodriguez and Vicente Chandler 1967)

Cane yield ($Mg\ ha^{-1}$)	Base saturation (%)	pH	Exchangeable bases ($mEq\ 100\ g^{-1}$)	Exchangeable Al^{3+} ($mEq\ 100\ g^{-1}$)
25	20	< 3.8	3	8
100[a]	58	> 4.8	8	2

[a] Limestone was applied up to 45 t ha^{-1} over a 7-year period.

Al^{3+} with a consequent improvement in yield and quality of cane. The authors recommended lime and phosphogypsum application at levels of 4:2 or 2:4 for maximum production of cane and sugar.

Another category of problem soils is the highly toxic acid sulphate soils (thionic fluvisols). These are known as cat clays and are found in coastal, deltaic or estuarine areas of the humid tropics. Their formation is associated (Harmsen 1954) with polysulphides of iron (FeS, FeS_2). After draining they become very acidic with pH values below 2. Dent (1986) has listed the following agronomic problems associated with growing crops in acid sulphate soils:

1. Toxicity of Al^{3+} and possibly also Fe^{2+}, Mn^{2+} and H^+ ions.
2. Low base status and nutrient deficiency.
3. Restricted root development and deformation.
4. Restricted microbial growth with reduced nutrient release.
5. Most crops are susceptible to diseases.

A detailed investigation of sugarcane cultivation in the acid sulphate soils of Guyana is presented by Evans (1965). These acid peat soils (pegasse) are formed by anaerobic decomposition of woody material. The soils are low base saturated with tertiary pyrites and Al^{3+} is the predominant cation. The distinctive features of cane grown in cat clays are stunted, coralloid roots and leaves show varying degrees of reddish freckling (Evan 1965). The factors limiting growth are the presence of high amounts of Al^{3+} (more than 60%) and Fe^{2+}, a low pH, low levels of nutrients including silicon, and the unavailability of P. Evans (1965) demonstrated convincingly that such soils can be reclaimed by leaching with seawater accompanied by a modest dressing of limestone. He obtained a satisfactory yield of about 125 Mg ha^{-1} as against 22–25 Mg ha^{-1} before treatment.

3.2 Climate

Sugarcane is a tropical plant and the cultivated varieties are all grown in warm countries (Simmonds 1986); the climatic boundary is further defined by the coldest mean monthly temperature isotherm of 18.3 °C (Cambell 1967). It is grown from sea level to 1500 m and between the latitudes 36.7 °N and 31.0 °S. But as the elevation increases, so the temperature decreases and the crop requires longer period for growth and development. The ideal climate is a long, warm growing season and a fairly dry, sunny, cool but frost-free ripening and harvest season, free from hurricanes and typhoons (Humbert 1968).

Climatic parameters like light, temperature and rainfall (moisture) have a profound influence on yield and quality. Clements et al. (1952) aptly stated, 'the size of the theoretical crop is determined by the energy available to it and its inherent ability to fit into the particular energy level'. The dependence of plant productivity on the environment is stressed by Boyer (1982)

Table 3.9. Climatic elements influencing yield and quality of sugarcane

Climatic elements	Processes affected
Air–CO_2 concentration	Alters rate of photosynthesis
–Density of pollutants	plant injury with possible reduction in yield and quality
–Ozone	Ultraviolet injury, growth retardation
Light–Day length	Influences flowering
–Intensity	Controls photosynthesis
Rainfall–Amount	Causes floods or drought
–Distribution	Requires special management
–Number of rainy days	Determines planting and harvesting seasons
Humidity	Desired at vegetative phase but hinders ripening and sugar accumulation. Affects evapotranspiration and accentuates incidence of disease
Temperature	
–Seasonal and daily fluctuations	Alters rate of photosynthesis and accumulation of photosynthate
–Low temperatures	Cold injury, low germination, reduced tillering
–High temperatures	Heat injury and water deficit
Wind	Lodging with poor-quality cane
Cyclones and hurricanes	Canes uprooted with heavy yield losses

and production invariably falls short of the potential. Hence, it is imperative to understand the plants' response to the environment to achieve maximum productivity. The climatic elements influencing the yield and quality of cane are listed in Table 3.9.

3.2.1 Carbon Dioxide Concentration

Ordinarily CO_2 is in short supply for photosynthesis (Clements 1964). However, the recent literature indicates, rather conclusively that the CO_2 concentration in the atmosphere has increased from ca. 280 ppm before the Industrial Revolution to the present level of $\sim$ 350 ppm (Adams et al. 1990). This may possibly double in the next 50 years due to the burning of fossil fuels. An increased CO_2 concentration may possibly increase the yield of several crops (Adams et al. 1990). But a distinct disadvantage of the continued build-up of CO_2 is likely to be a rise in the surface air temperature by 1.5–5.5 °C; this is termed the 'greenhouse' effect and will lead to a consequent increase in drought and aridity. Estimates show that a doubling of the CO_2 concentration leads to an increased heating of 4 $W\,m^{-2}$ (Hume and Cattle 1990).

Sugarcane benefits from an increased atmospheric CO_2 concentration. Hartt and Burr (1965) demonstrated an increased photosynthetic rate as CO_2 concentration was increased from 0.01 to 0.07% but this became saturated at 0.06%. Varietal differences were also observed; H37-1933 was saturated at 0.06% CO_2, whereas H38-2915 was saturated at 0.03%.

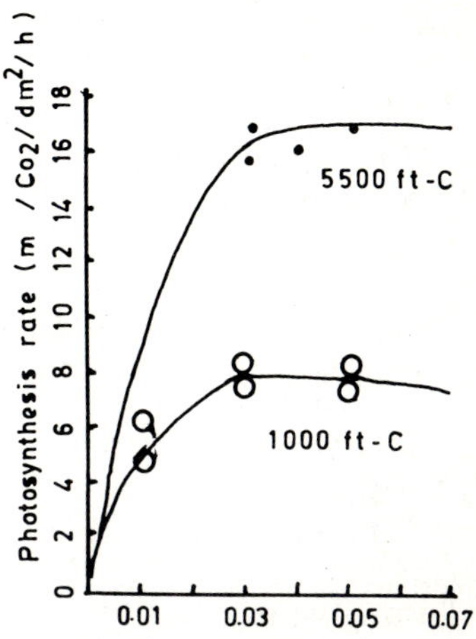

Fig. 3.3. Effect of CO_2 concentration on photosynthetic rate at 1000 and 5500 ft-C, cv. H38-2915. (Hartt and Burr 1965)

A strong interaction was found between sunlight and CO_2 (Hartt and Burr 1965), and with increased sunlight CO_2 fixation in cane was enhanced (Fig. 3.3). Jones (1985) concluded that the quantum efficiency of light for CO_2 fixation by C_3 and C_4 plants remains fairly constant at 25 °C (i.e. 0.05–0.06 mol CO_2/mol photons). But above 25 °C, C_4 grasses were more efficient.

3.2.2 Pollutants

Major pollutants in the air are: SO_2, O_3, NO_x, HF and NH_3. These cause a reduction in the yield of several grasses and legumes. This author has no information on the effect of pollutants on yield and quality of cane, but Davies (1990) reasoned that grasses are more sensitive to SO_2 pollution even at 11 parts per billion (ppb). The effect was more pronounced under low irradiance (40 W m^{-2}) and short photo-periods than under high irradiance (130 W m^{-2}). It was concluded by Davison and Bailey (1982) that SO_2 pollution reduced the resistance of rye grass to cold injury, possibly due to the plant's inability to detoxify the pollutants under conditions of low irradiance.

Ozone also depresses photosynthesis. Ormord et al. (1981) found that the addition of ozone at 60–150 ppb to an atmosphere containing 40 ppb SO_2 led to a considerable decrease in photosynthesis by field beans. Furthermore, the combined effect of O_3 and SO_2 was more pronounced than the effect of either of the gases alone.

3.2.3 Light

C_4 grasses normally grow under conditions of higher solar radiation than C_3 plants (Jones 1985) and a comparison of photosynthesis between C_3 and C_4

plants indicates that the leaves of C_3 plants were light saturated at 35–70 mW cm^{-2} compared with 105–140 mW cm^{-2} in C_4 plants (Bull 1969). The ability of C_4 plants to utilize higher light intensities than C_3 plants is associated with a greater affinity for CO_2 at the carboxylation site (Bull 1969). Clements (1940) demonstrated rather convincingly that the marked differences in cane yield at two locations in Hawaii were essentially due to differences in sunlight, and doubling the number of hours of sunshine more than doubled the yield. Similar observations were made by Irvine (1967) who observed nearly 25 times more photosynthesis when the light intensity increased from 2150 to129 000 lx. However, he emphasized that in addition to light, leaf thickness and porosity contributed towards increased photosynthesis in some cane cultivars.

Varietal differences were also significant in respect of light utilization (Hartt and Burr 1965). In general, most cane varieties showed maximum photosynthetic rates at 60 000–80 000 lx. But H52-123 was saturated at 100 000 lx, whilst H45-1045 was not light saturated even at 140 000 lx. Hartt and Burr (1965) indicated that reduced photosynthesis was due to a lack of translocation and accumulation of photosynthate in the leaves (product repression). Hence the translocation of photosynthate from the primary source (i.e. the leaf) is of the utmost importance for continued photosynthesis at the maximum rate.

3.2.4 Frost Stress

Freezing conditions are met in Australia, Argentina, Egypt, Florida, Louisiana, Iran, North India, Pakistan, South Africa and Zimbabwe. In sugarcane, chilling temperatures for long periods cause restricted growth, chlorosis, apparent P deficiency and poor sett germination/stubble sprouts. The severity of the frost is related to soil moisture and damage is greater during dry spells with no protective cover. Frost injury is accentuated with high N content in tissues and eventual tissue death. The earliest work on cold injury in sugarcane seems to be that by Faris (1926) in Cuba. White chlorotic bands were noticed following cold injury, and these were locally known as 'manchas blancas' or 'Faris bands'. Young leaves were affected more and cold chlorosis could occur even at 10 °C in sensitive varieties like H109 and D1135.

Irvine (1968) tested 50 clones of *Saccharum* and showed that *S. officinarum*, *S. robustum* and *S. edule* were least resistant to cold (− 3 to − 4 °C). However, varieties of *S. spontaneum*, *S. sinense* and interspecific hybrids exhibited greater tolerance. The most outstanding varieties were CP57-526 and NCO310 which showed least tissue damage. The most severely damaged variety was CP47-193 (Irvine 1968). A summary of stalk weights and quality parameters following 92 days under a freezing temperature is given in Table 3.10 (Irvine 1968).

There was no significant change in stalk weight following a 3-month period of freezing but a significant improvement is seen in the sucrose and juice purity with reduced acidity (Irvine 1968).

Table 3.10. Effect of freezing temperature (– 4.4 °C) on quality parameters (mean of 92 days after freezing). (Irvine 1968)

Stalk weight (kg)	Crusher juice analysis			
	Sucrose (%)	Purity (%)	Acidity (ml 0.1 *N* NaOH/10 ml)	Gums (mg/10 ml)
1.12	13.61	84.51	2.00	2.12

The purplish tips and margins of young leaves noticed in North India were attributed to the accumulation of anthocyanins following P deficiency under chilling temperatures.

Reviewing the work on C_4 grasses, Jones (1985) concluded that cold hardiness in genotypes could possibly be increased by higher concentrations of: (1) carbohydrates; (2) proteins; and (3) unsaturated short-chain lipids. The solution rests with the development of tailor-made varieties for cold resistance. In the Punjab (subtropical North India) a polyethylene mulch is spread to protect the sprouts. Irrigation is an oversimplified method to control cold injury. Overhead sprinkling provides effective frost protection but the investment cost is high.

3.2.5 Temperature

This climatic parameter is paramount in influencing the growth and development of cane. In the sugarcane belt, temperatures range from freezing to over 50 °C. At high temperatures above 50 °C, C_4 plants may produce heat shock proteins (Jones 1985).

Temperature effects are more pronounced on germination, stalk length and diameter with the optimum temperature being 30–32 °C. Humbert (1968) observed no growth at 1°C and limited growth at 21 °C. Leaf area expansion is related more to air temperature and is optimal at 22 °C (Bull and Glasziou 1963).

It is difficult to separate the effects of ambient and soil temperatures on the growth of cane. Clements and Nakata (1965) observed 10 °C as the critical soil temperature for germination. Interaction between the air, root temperature and clones was referred to by Brodie et al. (1969). The growth of 57-5174 was outstanding at the lowest root temperature (12.76 °C). Lower temperatures were instrumental in enhancing sucrose levels in stalks by decreasing reducing sugars. In general, dry matter and stalk elongation increased with an increase in air temperature from 17.2 to 22.2 °C and with an increase in root temperature from 12.76 to 21.1 °C. The highest amount of dry matter and greatest stalk elongation were noticed in a warm house (air temperature 22.2 °C) where the root temperature was 21.1 °C.

In Hawaii, Mongelard and Mimura (1971) maintained that low air and root temperatures caused a decrease in dry matter production and that this was associated more with a decrease in leaf growth under the influence of low air (15 °C) and root temperatures (13.5 °C).

Photosynthesis was equally affected by air and root temperatures. Burr et al. (1957) observed a reduction of 84% in photosynthesis of cane as the ambient temperature decreased from 23.1 to 13.6 °C. Similarly, an 83% decrease in photosynthesis and a sevenfold decrease in ^{32}P uptake was noticed in cane as the root temperature decreased from 22.1 to 16.7 °C. Hartt and Burr (1965) confirmed these trends. At 32.2 °C, ^{14}C counts were 185 as compared with 148 counts at a lower air temperature of 25.6 °C.

In general, a reduction in photosynthesis and nutrient uptake takes place following decreased root or air temperatures and the effects are additive. According to Srivastava and Johari (1979) cane production is more related to the coefficient of variation in the mean air temperature. But Garza (1968) concluded that under Mexican conditions, the accumulated mean temperature is above 10 000 for optimum sucrose synthesis. During sugar accumulation a low temperature seems essential for higher sucrose production.

Das (1933) proposed the day–degree concept, which is defined as the temperature above an average maximum of 70 °F. He has shown a close relationship between cane yield and the total effective warmth received. For a 1-year crop, accumulated day-degrees ranged from 5700 to 93 000.

Emphasis is now placed on the modified growing degree days (MGDD):

$$\mathrm{MGDD} = \left[\left(\frac{T_{max} + T_{min}}{2}\right) - 10.0\right] \mathrm{daily}.$$

MGDD gives the 'effective degree' index and any daily minimum temperature below 10 °C is equal to 10 °C and any daily maximum temperature above 32 °C is equal to 32 °C. It was found that the emergence of shoots requires 150 degree days.

The total MGDD in some important cane-growing countries for a 12-month crop ranged from over 2000 in South Africa, Peru, Argentina and Spain to more than 6000 in Trinidad, Thailand and the Philippines. It is inferred that radiation coupled with temperature influences the growth of cane and is congruent with the observations of Venkataramana et al. (1984).

3.2.6 Rainfall and Humidity

Sugarcane is grown in areas receiving rainfall of 500–2500 mm per annum. Regions of low and erratic rainfall are irrigated. A very humid environment accentuates fungal diseases (Blackburn 1984). Rainfall determines the planting and harvesting periods and a high humidity drastically reduces the net sugar recovery. A detailed investigation by Venkataramana et al. (1984) on the effect of weather parameters on the growth pattern of sugarcane varieties

Table 3.11. Coefficients of correlation between some weather parameters and leaf area for two sugarcane varieties grown under conditions of normal irrigation. (After Venkataramana et al. 1984)

Variety	Air temp.	Relative humidity	Rain-fall	Rainy days	Soil temp.	Mean evaporation rate
CO419	– 0.864**	0.909**	0.678*	0.687*	– 0.920**	– 0.779*
CO1148	– 0.624^{ns}	0.736*	0.631^{ns}	0.794*	– 0.704**	– 0.594^{ns}

* Significant at 5%; ** significant at 1%; ns, not significant

revealed that leaf area production was positively correlated with relative humidity, rainfall and the number of rainy days but was negatively correlated with the mean ambient and soil temperatures and evaporation rates (Table 3.11).

3.2.7 Other Climatic Parameters

Wind has both direct and indirect influences on sugarcane. Wind-damaged cane has reduced rates of photosynthesis and translocation of photosynthate. Tropical cyclones (hurricanes and typhoons) occur in the Caribbean, Indian Ocean and China Sea and under their influence frequent cane damage is seen in Bangladesh, Mauritius, Fiji and Australia. Sugarcane in Jamaica suffers damage from hurricanes, and typhoons occur frequently in the Philippines.

The microclimate is defined as one where climatic factors like temperature and humidity inside the crop canopy differ from conditions outside. Variation in the microclimate occurs due to increased humidity and decreased air circulation. The temperature inside the crop is 5–6 °C lower than outside (Parthasarathy 1972). Based on in-depth studies, Jen-hu Chang (1961) conceded that sugarcane, as a tall plant, has a distinct microclimate. High canopy roughness is responsible for high rates of evapotranspiration (ET). Albedo or reflectivity is much higher under plant cover than bare soil and increases with the age of the crop. The energy budget for a mature cane field is: 16% reflected radiation, 17% back radiation and 67% net radiation. Furthermore, 82% of net radiation or 55% of incoming radiation is consumed in evapotranspiration (ET).

Chapter 4 Seed Bed Preparation, Methods of Planting, After-Care and Cropping Systems

Conventionally the soil is prepared by a primary tillage consisting of one to two deep ploughings with mould board ploughs or chisels to uproot the old stubbles and break the land into clods. This is followed by a secondary tillage with two to three harrowings to break the clods and bring the soil to a fine tilth. If planting is done in a system of ridges and furrows then a fine tilth is not so necessary. But for flat-bed planting a good tilth seems essential. Sugarcane, however, does not require as fine a tilth as that needed for small-grain crops. In Cuba and Brazil deep cultivation to a depth of at least 30 cm is recommended to break the subsurface hard pans and decrease the bulk density. In Hawaii subsoiling is practised (Husz 1972) to a depth of 45–60 cm. In Australia and Taiwan intensive cultivations are practised to break the surface crust, hard pans and ridges. Husz (1972) suggested that an approximate pore ratio of 1:1:1 for large, medium and small pores assures the establishment of both plant and ratoon canes. In areas of high rainfall and low-lying flat land requiring adequate drainage, cane is planted on cambered beds, as in Guyana and Fiji (see Chapter 6). These beds are frequently 6.1 m wide with side drains of 0.6 m width and 0.45 m depth. In the Louisiana bank system, cane is grown on special ridges and furrows known as 'turtleback'. The ridges are 0.45 m high and spaced at intervals of 1.8 m (see Chapter 6). The special feature of this system is that it permits the total mechanization of cane culture (Blackburn 1984). On sloping land ridges and furrows are made along the contours and the cane is planted in the furrow and covered by breaking one ridge. Another unbroken ridge conserves soil and moisture and prevents soil erosion (loc. cit.).

Cultural practices for all crops, including sugarcane, develop through native wisdom and the ingenuity of the farmers who are familiar with the local problems. In the early days of cultivation, land had to be cleared of bushes, shrubs, trees, etc. and all sorts of implements were used. The pressure on the land was minimal. As this pressure became more acute the farmer had to settle for intensive cultivation.

It is only in recent times that serious efforts have been made to discover the optimum number of possible cultivations for particular crops. Deep cultivation with heavy ploughs, chiselling and subsequent harrowing to break the clods became a regular practice to obtain a fine tilth. In the meantime, more potent herbicides were produced and arrived on the market. Hence, a rethinking of methods appears to be even more necessary. Research in Brazil and Cuba amply demonstrated the need to plough the field to a depth of at least 30 cm, where the root zone of sugarcane is active.

Fernandes and his associates (1981) concluded that the soil profile at 40–50 cm deserves special attention because a maximum number of roots are found in this layer. Ploughing to a depth of 45 cm is all the more necessary since moisture retention is maximal here and the bulk density can be maintained at 1.2–1.3 $g\,cm^{-3}$.

4.1 Conventional vs. Minimum Tillage

The high cost of production and an acute shortage of labour led to a rethinking on the conventional practice of deep tillage for sugarcane. Research results came out in favour of minimum tillage in normal soil for sugarcane plants and subsequent ratoons. Sanchez (1976) stated that minimum or no tillage is an ancient practice in the tropics. Referring to the work of Vicente Chandler in Puerto Rico, he showed that there were no differences in cane yield between practising tillage and no tillage (Table 4.1).

However, conservation tillage assumes a greater importance in rain-fed sugarcane where the conservation of soil moisture is a prerequisite. Glaz and his associates (1989) clearly showed that a minimum tillage of two shallow cultivations is enough to realize a high yield of cane and sugar in the Histisols of Florida (Table 4.2).

Similar observations were made by Ricaud (1971) who stated that two to three cultivations are required for plant and stubble crops of cane grown in the silty or clay soils of Louisiana.

Workers in South Africa have strongly advocated minimum tillage for plants and subsequent ratoons (Moberly 1969, 1972; Iggo and Moberly

Table 4.1. Sugarcane yield in tilled and untilled soils

	Catalina clay		Mucara clay	
	Tilled	Untilled	Tilled	Untilled
Sugarcane yield ($Mg\,ha^{-1}$)	111	109	66	80

Table 4.2. Effect of number of shallow cultivations on cane and sugar yield ($Mg\,ha^{-1}$). (Glaz et al. 1989)

Low: no cultivations		Medium: 2 shallow cultivations		Med.-high: 3–4 shallow cultivations		High: 6–8 shallow cultivations	
Cane	Sugar	Cane	Sugar	Cane	Sugar	Cane	Sugar
84.8	9.28	91.4	9.68	89.9	9.75	95.6	10.38

Table 4.3. Effect of tillage practice on cane yield and root distribution. (McIntyre and Barbe 1989)

		Minimum tillage (MTP)		Normal tillage (NP)	
Trial No.	Variety	Cane yield ($Mg\,ha^{-1}$)	No. of roots/100 cm^2	Cane yield ($Mg\,ha^{-1}$)	No. of roots/100 cm^2
1	B51129	80.3	2.8	80.0	2.9
1	Trion	83.3	2.1	75.9	2.2
2	M377/56	76.0	2.7	74.0	2.4
3	M377/56	94.0	2.4	95.7	2.6
4	M574/62	60.1	2.4	59.0	2.2
5	M555/60	45.2	2.0	49.0	2.7

Soil type is dark magnesium clay

1976). Moberly (1969, 1972) observed that a conventional plough depth of 20–25 cm is quite adequate and pleaded for minimum tillage through chemical ploughing using glyphosate. Hadlow and Millard (1981) applied glyphosate at 2.8–3.5 kg active ingredient (ai) ha^{-1}, and if no bush clearing or land levelling is desired then glyphosate can be sprayed at 1.48 $kg\,ha^{-1}$. The incorporation of lime in acid soils under minimum tillage conditions has been attempted by Meyer and co-workers (1989). In Mauritius, minimum tillage is an accepted practice on sloping and erosion-prone lands (McIntyre and Barbe 1989). In normal practice (NP), cane is replanted on flat land by mechanically uprooting the stools by using a subsoiler (40 cm depth), followed by disc harrowing and then the furrows are opened for planting. In minimum tillage (MTP) old cane stools and weeds are killed by an application of Round up (glyphosate) and then the furrows are opened for replanting. McIntyre and Barbe (1989) reiterate that minimum tillage is an acceptable practice in all soil types except in heavy, dark, magnesium clays. The root distribution is more concentrated at the 30-cm layer in MTP compared with deeper root extension in NP (Table 4.3).

4.2 Planting Methods

Basically there are two methods of planting, namely, manual and mechanical. In manual planting several variants are available depending on the agro-ecological and socio-economic conditions. The flat method of planting is essentially to conserve soil moisture since cane is grown under rainfed conditions or receives supplemental irrigation. The land is prepared in a fine tilth by one or two deep ploughings. Shallow furrows of 10–15 cm depth are opened with a wooden plough and the setts are placed end to end (or overlapping) in the furrow. They are covered with soil immediately and this is compacted with a heavy wooden plank. In this way soil moisture is conserved and germination depends on the sett moisture; it can take a pretty long time (up to 2 months).

Trench planting is also adopted to save water. After the land has been thoroughly prepared trenches are made at a distance of 1–2 m. Each trench is approximately 20 cm deep and 40 cm wide. About 2–4 basketfuls of farmyard manure or compost are added. In the loosened soil plus manure mixture the setts are planted flat. The chief advantage is that the cane gets a good anchorage and does not lodge. A saving of about 15% irrigation water and a yield increase of about 5–10 $Mg\,ha^{-1}$ can be anticipated over flat/shallow planting. The greatest disadvantage is that it is highly labour intensive. The method is used in some parts of India and Indonesia where labour is relatively cheap. Blackburn (1984) has referred to planting in cane holes in the drier regions of Barbados, Mauritius, Réunion, parts of Jamaica and the sloping land of Antigua but this method is now obsolete. In areas of high rainfall and clay soils with water-logged conditions (see Chapter 6) slant planting (at an angle of 45°) is adopted (in India this is known as the Partha method after the late Dr. S. V. Parthasarathy). In this technique, three bud setts are planted (2–5 cm deep), either on the top of the ridge or on the side of the ridge, inclined at an angle of 45° so that one bud is buried in the soil and two buds are exposed. The two exposed buds germinate. When the monsoons recede the sett is pressed horizontally so that one or two germinating shoots from each sett strike roots. At this stage fertilizers are applied.

A new planting technique, known as IISR 86206 (later IISR 8626), was developed by Panje (1970) and this was supposed to be universally applicable. The five-digit code refers to: (1) planting vertically in a 50-cm-deep, 20-cm-wide pit; (2) the pit is filled full with soil, manure and fertilizer; (3) trenches are irrigated; (4) trenches are spaced 90 cm apart; and (5) 45-cm-long top setts are pregerminated into shoots. The yields were high and the inputs were efficiently used but commercial adoption was poor and hence the technique became of academic interest only.

A modified version of the IISR 86206 method is the spaced transplanting technique (STP) developed by the Indian Institute of Sugarcane Research (IISR) in Lucknow. This technique is well suited for achieving a reduction in seed rate by one-third and it increases the multiplication ratio of buds. A settling nursery of 50 m^2 (5 ×10 m) consists of small plots of 1 m^2. Well-rotted manure (if available) and 1 kg gamma BHC (benzene hexachloride, γ-isomer or lindane) are applied. A single bud sett treated with 0.1% aretan (mercury chloride) is planted vertically with the bud on top. About 30 000 single-bud setts are planted vertically in the 50 m^2 area. These buds are covered with cane trash or paddy straw to protect them from the hot sun. Watering of the nursery is done by rose can. The settlings are ready for transplanting to the main field after 3–4 weeks. The STP technique can be adopted for either trench or flat planting. The spacing is 90 cm between rows and 60 cm between transplants. Gap filling, earthing up and fertilization are carried out 4–6 weeks after the settlings have been transplanted to the main field. STP is specially suited for seed multiplication of newly released varieties. This system produces many tall stalks, an uniform stand and there is better utilization of light and other resources. The crop is not prone to lodging. The anticipated yield is over 150 $Mg\,ha^{-1}$, which constitutes an increase of 20–25% over the ridge and furrow system.

The common manual method of planting in Asia, Africa and some countries of South America is the ridge and furrow system. The ridges and furrows are spaced 0.9–1.0 m apart and the furrow length is 10–15 m, depending on the slope of the land. Irrigation channels are formed crosswise at a distance of 20–25 m. Planting of two or three eye-bud setts is made in either a wet furrow or a dry furrow. However, immediate irrigation is necessary when planting is in a dry furrow. If heavy rains are expected, the setts are planted on the top of the ridge or half way between the ridge and furrow.

In any manual method the depth of planting is 2.5–5.0 cm and the setts are covered with soil to prevent dryage. Germination essentially depends on the sett moisture and total sugar content.

The most innovative method of manual planting is the ring planting technique which has been well described by Yadava (1991). This method encourages development of synchronous mother shoots. The field is marked at regular intervals of 120 cm. With the help of a long rope, straight lines are drawn both ways and at the point of intersection a ring of 90 cm diameter and 45 cm depth is made. A space of 30 cm is left between the two rings (Fig. 4.1a). Each pit receives 5–8 kg farmyard manure or compost, 5 g aldrin (10%) or BHC (10%), 40 g urea and 10 g each of super phosphate and muriate of potash. Healthy setts with three eye buds, preferably obtained from heat-treated cane, are chosen. Twenty such setts are arranged horizontally in a circle in each pit (Fig. 4.1b). Thus, each pit has 60 eye buds and about 25–35 millable canes are formed in each pit. There is no scope for tillering. Only mother shoots are formed and these are heavier and thicker. Nearly 180–200 $Mg\,ha^{-1}$ cane can be expected and this has a better juice quality. A tractor-mounted pit digger for ring planting has been developed at IISR, Lucknow (Sharma and Singh 1988) and this can make 500 pits per 8-h working day.

In the developed countries of the United States and Australia and in many countries of Asia, Africa, Latin America and the Caribbean manual planting has been totally dispensed with due to non-availability and/or very expensive labour costs. Cane planters have been developed to suit local conditions. An automatic tractor-drawn (35 hp) sugarcane planter has been developed at IISR, Lucknow (Sharma and Singh 1988). The planter is an attachment to the standard hydraulically mounted tool frame with two ridger bodies (Fig. 4.2). The machine consists of an automatic seed cane metering mechanism, a seed chute, a gamma BHC tank, a fertilizer application unit, a covering device and a tamping roller. The automatic metering mechanism consists of a hopper, a pusher mechanism, a weighted flap and a picker unit. The design is such that even bent cane pieces can be metered without any problem.

It is capable of planting 3.5 ha per 8-h working day with four persons working the machine.

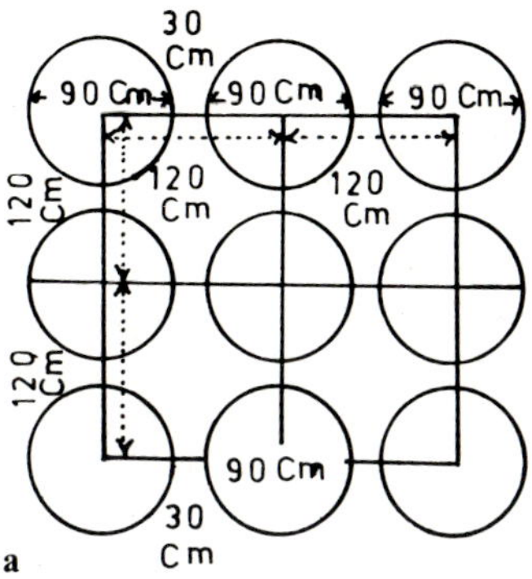

a

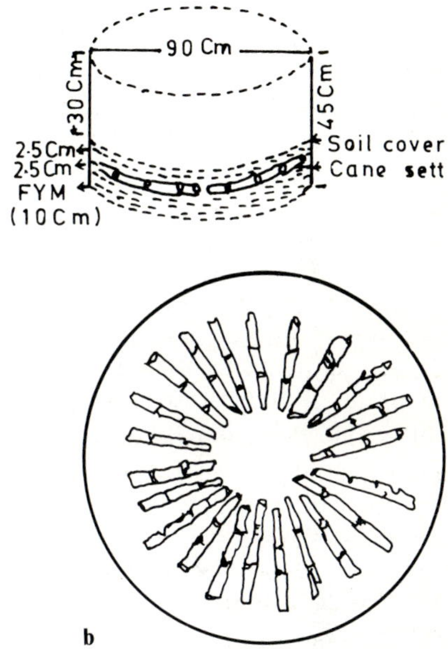

b

Fig. 4.1. **a** Ring method of planting **b** Sett arrangement in a pit. *FYM*, farm yard manure (Yadava 1991).

Fig. 4.2. Automatic tractor-drawn sugarcane planter in operation. (Sharma and Singh 1988)

The IISR has also developed a power-operated sett cutting machine which can cut sugarcane stalks into three- or four-eye-bud seed pieces (Sharma and Singh 1988). This machine consists of 400-mm-diameter circular serrated saw blades mounted on a mild steel iron framework. Four persons are required to work it and two persons hold four to five entire canes and push these against the rotating (700–800 rpm) saw blades. The setts of predetermined length are cut, roll over the slanted wooden platform and drop into a tank of fungicidal solution. The output is 12000–13000 three-eye-bud setts per hour and bud damage is in the order of 1–2%.

4.3 Spacings and After-Care

Irvine and his associates (1980a) have provided an admirable review of the historical and theoretical aspects of spacing in sugarcane. They discerned that in natural stands wild canes (*Saccharum spontaneum*) grew as close as 2–5 cm, and wider spacings adopted for commercial varieties are occasioned by practical considerations. In the early part of the eighteenth century, the spacing adopted for commercial crops varied from 45 to 60 cm in the United States and South America. Even now, inter-row spacings of 69–90 cm are adopted in many parts of Asia and Africa where man and animal power are used.

It is therefore logical to assume that wider spacings were necessitated by mechanized cultivation. Thompson and du-Toit (1965) reasoned that the long-cycle canes of 18–20 months require wider spacing. But Bull (1975) observed a significant yield increase in both plant and ratoon canes when the inter-row spacing was reduced from 140 to 50 cm. A very large response to reduced spacing was seen in cultivars H51 8029 and H1695-1 and this is attributed to the erect leaves and associated high leaf area index (LAI). Bull even pleaded that a high yield potential can be realized if clonal selections are made under conditions of closer spacing. Similar observations were made by Matherne (1972) in Louisiana where closer inter-row spacings were superior to wider spacings. He obtained the highest yields of 107 and 113 $Mg\,ha^{-1}$ in a close spacing of 90 cm with varieties L60-25 and CP61-37, respectively. A wider row spacing of 180 cm gave the lowest yield in all the varieties tested.

An effort was made to fit an empirical model with yield as dependent and spacing regime (population) as independent variable. Normally, the plant population in cane varies from about 14 to 20 m^{-2} in accordance with the available spacing. The yield trends for different populations are fitted to an asymptotic curve, given as

$$y = \frac{ap}{1 + bp + cp^2},$$

where y = yield response, p = population variable, and a, b, c are constants.

Rearranging the equation:

$$1/y = (b/a) + (c/a)p + (1/a)(1/p).$$

The fitted inverse function of y for variable p by using an asymptote is given as

$$1/y = 0.042411 - 0.0001p - 3.17498/p.$$

The estimated standard error for 1/y was ± 0.000874. It is reasonable to establish an asymptotic relation between yield and plant density. However, Thornley and Johnson (1990) argue that parabolic and square root models explain the yield and plant density relation better than the other models.

Cane yield is a function of stalk population and height and weight per cane, but the stalk population accounts for nearly 70% of variation in cane yield (Hunsigi and Srivastava 1975). Irvine and Benda (1980) concluded that a higher yield under closer spacing is due to an exponential increase in stalk population despite a linear weight decrease. Further studies led them to infer that a narrower intra-row spacing of 19 cm gave the highest yield (91 Mg ha^{-1}) when the rows were spaced at 183 cm. They averred that close spacing in a short growing season resulted in the highest cane yield despite a reduction in weight or height per cane (Table 4.4).

A decrease in diameter with crowding is obvious. A high cane yield from close spacing is also associated with a high biomass production.

The sugar content was not influenced by spacing regimes but a higher sugar yield per hectare under conditions of narrower spacing is essentially due to the higher net cane yield. Irvine et al. (1980b) reiterate that the production technology for planting narrow rows would open new vistas in achieving higher yields, even under highly mechanized cultivation. They tested three whole-stalk harvesters (Cameco, J and L, and Thompson) and three cut-chop harvesters (Cameco, Class and J and L). The cut-chop harvesters performed better on 61-cm rows than did the whole-stalk harvesters. The adoption of narrow spacing rests eventually in the development of

Table 4.4. Effect of row spacing on yield, yield components and cane quality (cv. CP65-357). (Modified in part from Irvine and Benda 1980)

	Row spacing (cm)			
Parameter	19	38	76	152
Shoots (10^3)	228.0	109.0	83.0	58.0
Biomass dry weight (Mg ha^{-1})	61.00	46.00	37.00	20.00
Net cane yield (Mg ha^{-1})	171.00	138.00	112.00	60.00
Sucrose (Mg ha^{-1})	16.50	12.00	9.00	4.70
Sucrose (% cane)	9.63	8.72	7.95	8.07
Fibre (% cane)	11.49	11.15	10.47	10.08
Stalk height (cm)	250.00	274.00	282.00	187.00
Stalk diameter (cm)	2.08	2.39	2.51	2.77
Stalk weight (kg)	0.82	1.65	1.79	1.41
LAI	8.20	6.50	4.80	3.30

appropriate technology under conditions of mechanized cultivation (Irvine et al. 1980b).

The serious objections to narrower spacing are the high costs of seed cane, cultural practices and harvesting. Freeman (1968) expressed the apprehension that the increased cost of cultivation may negate the gains achieved under narrower spacings. Rough estimates show that 2 t seed per hectare may be required in wider row spacings (1.8 m) compared with 6 t ha^{-1} in closer row spacings (0.6 m). It is prudent to use two- or three-eye-bud pieces since the whole cane, when planted, will not germinate fully due to apical dominance.

On the whole, the early and short duration varieties respond well to closer spacing. Sundara and Subramanian (1989) demonstrated that the short-duration variety CO A7601 gave a higher yield with a narrower spacing of 60 cm. This is attributed to a high stalk density in both plant and ratoon canes (Table 4.5). Similarly, this author has obtained a high cane yield of an early maturing variety, i.e. CO C671, grown in clay soil (Vertisols) under conditions of narrower spacing of 60 cm.

In contrast, workers in South Africa contest that narrower spacings result in higher yields under all conditions. They observe that dense stands do generate a high plant population at the initial stages but that tiller mortality is equally high as a consequence of within-row shading. Excessive mortality is also associated with moisture stress in closely planted crops. Similar apprehension that moisture stress may be limiting with closer spacing and may result in greater tiller mortality was expressed by Hebert et al. (1965) in Louisiana. It is proposed that, wherever feasible, inter-row spacings should be reduced to 0.6–1.0 m and that implements like planters, harvesters, etc. need to be developed to exploit the full potential of the crop. Clonal selections should also be made under conditions of narrower spacing.

The spacings adopted in the sugarcane-producing areas of the world under commercial cultivation are given in Table 4.6.

After-care consists of two or three inter-cultivations to break the soil crust and allow the easy flow of water. In Mexico, 'discostre' or cross harrowing is carried out to break the crust. This is particularly effective on soils that shrink on drying and crust badly. In Taiwan, traditional inter-tillage was

Table 4.5. Cane yield of the short duration variety COA7601 as influenced by inter-row spacing. (Modified in part from Sundara and Subramanian 1989)

	Yield ($Mg\,ha^{-1}$)		Stalk density ($\times 10^3$)		Pol (%)		Sugar yield ($Mg\,ha^{-1}$)	
	S1 (90 cm)	S2 (60 cm)	S1 (90 cm)	S2 (60 cm)	S1 (90 cm)	S2 (60 cm)	S1 (90 cm)	S2 (60 cm)
Plant crop	71.91	81.61	89.44	119.12	16.40	16.62	9.06	10.31
Ratoon I	75.12	86.06	83.99	92.09	16.20	16.27	8.40	9.67
Ratoon II	47.55	49.75	49.26	56.54	15.16	15.16	4.96	5.04

Table 4.6. Row spacings for sugarcane in different countries. (Blackburn 1984)

Country	Spacing adopted for commercial cultivation (m)
Australia	1.4
Brazil	1.2–1.5
Indonesia	1.2–1.5
Puerto Rico	1.2–1.35
India	0.6 (short and early [8–10 month crop] duration varieties) 0.9 (Mid-late and late [14–18 month crop] varieties)
Swaziland	1.4
Mauritius	1.5
Trinidad	1.5
South Africa	1.5–1.8
Jamaica	1.2–1.8
Argentina	1.6
Cuba	1.65
Taiwan	1.20
Hawaii	1.5–2.4
Philippines	1.25
USA (Louisiana)	1.8
USA (Florida)	1.5
Zimbabwe	2.0

done by hand hoeing to control the weeds. But more recently mechanical methods have been used to break the ridges and bank up cane rows. This provides better anchorage for the standing cane.

4.4 Intercropping and Cropping Systems

Intercropping is defined as growing two or more crops simultaneously on the same area. The crops are not necessarily sown at exactly the same time and their harvest times may also differ, but they are usually 'simultaneous' for a significant part of the growing period (Willey 1979). Crop intensity may be considered in both time and space dimensions. There is intercrop competition during all or part of the crop growing season. The advantages of intercropping are many and include control of pests, diseases and weeds and, above all, yield stability. This technique is essentially used by small farmers in developing countries. Sugarcane can accommodate one or two rows of short-duration cereal, legume and oilseed crops to increase the income of the farmer. In sugarcane intercropping an 'additive series' (not a 'replacement series') is used where one or two lines of another crop are grown in addition to the main crop.

Sugarcane is usually the 'dominant' species when prostrate plants are taken as the intercrop. Mutual co-operation has been observed when cane is planted with one or two rows of groundnut or soyabeans and the yield of each species is greater than expected. The plausible explanation advanced is that the addition of N by the legume benefits the component crop (sugarcane). Willey (1979) stated emphatically that the top canopy should consist of a component crop with a high light requirement while the bottom component should have a low light requirement. The obvious reference is to a tall C_4 plant like sugarcane and a short C_3 plant like soyabean or groundnut.

Our studies have shown that the soyabean cv. Monetta is ideally suited as an intercrop in sugarcane and has a competitive index of less than unity (Fig. 4.3). The aggressivity value was negative, suggesting that soyabean was 'dominated' by sugarcane and had no adverse effect on the yield and quality of cane. The candidate species suitable as intercrops with sugarcane are: wheat, potatoes, peas, gram, finger millet, onions, garlic, coriander, radish, sugarbeet, soyabean, mustard and tobacco. Land equivalent ratios (LER) with complementary crops like beans, cowpeas, potatoes, peas and soyabean range from 1.5 to 2.0. Enahari et al. (1989) have shown that under Moroccan conditions, the ideal intercrops are broad beans, onions, leeks and potatoes. The reduction in cane yield due to intercropping was, on average, 6%, except in sugar beet where the yield reduction was 19% (Table 4.7). The least reduction in cane yield was for sugarcane with onion intercropping, followed by cane and a potato intercrop. Employing a radio tracer technique, Chang et al. (1969) observed serious competition for the nutrients P and K (band placed) in sugarcane and sweet potato. However, such competition was not

Fig. 4.3. Soyabean (cv. Monetta) intercropped in sugarcane

Table 4.7. Summary of sugarcane intercropping in Morocco. (Enahari et al. 1989)

Crop	Intercrop yield ($Mg\,ha^{-1}$)	Cane yield ($Mg\,ha^{-1}$)	Intercropped cane yield ($Mg\,ha^{-1}$)	Difference (%)
Broad bean	17.6	89.6	78.4	− 12.5
Onion	28.1	109.9	108.4	− 1.4
Leek	13.2	109.9	101.0	− 8.1
Potato	14.2	114.5	110.9	− 3.2
Mean	—	106.0	99.7	− 6.0
Sugarbeet	58.8	89.6	72.6	− 19.0

observed for cane and groundnut due to the over powering effect of sugarcane (dominant species).

In general, the crops used for intercropping in sugarcane (or in any crop for that matter) should be tailor-made to achieve complementarity as evidenced by a high yield in both the main and component crops. The best examples are the Monetta variety of soyabean and the early short-duration variety of pigeonpea (T-21), both of which are admirably suited as intercrops in sugarcane.

A sugarcane-based cropping system has a distinct advantage since sugarcane leaves large crop residues, besides being heavily fertilized in many cane-producing areas. Our studies on the red, sandy loam soils of Mandya have shown that paddy, cowpea and finger millet can exploit the residual fertility and give an optimum yield at half the recommended dose if harvested after sugarcane. Early maturing and short duration varieties have opened new vistas in cane-based cropping systems. It has already been pointed out that three crops of cane (one plant crop and two ratoon crops) can be grown in just 24 months with a production potential of 8–10 and 0.8–1.0 $Mg\,ha^{-1}\,month^{-1}$ for cane and sugar, respectively.

In India, sugar-rich short-duration varieties (8 months) have been evolved to fit into the multiple cropping system. Such systems require higher energy inputs than the conventional system (plant cane and first ratoon). In a detailed investigation, Sundara and Subramanian (1987, 1989) showed that out of five crop sequences, a sequence of short-duration cane (COA7601), first ratoon, finger millet and cotton gave the highest benefit/cost ratio of 3.13. This crop pattern gave yields of 195.6, 3.8 and 4.6 $Mg\,ha^{-1}$ for cane, finger millet and seed cotton, respectively (Table 4.8). The authors conclude that a cropping system based on short-duration sugarcane varieties requires 40% more energy input than the conventional system. The energy input through fertilizers is higher in short-duration cane sequences and 30–50% more nutrients (N, P and K) are absorbed than in the conventional system. Nevertheless, a system based on short-duration canes gave a 14–27% higher energy output than the conventional system (Table 4.8).

It has been observed that some sort of rotation is followed in cane-producing areas. In Louisiana (USA) and Mauritius, cane is rotated with leguminous crops. In Java, the land tenure system enforces rotation. In

Table 4.8. Effect of a sugarcane-based cropping system on the economic and biomass yield, energy output and benefit/cost ratio. (Modified in part from Sundara and Subramanian 1987, 1989)

Crop sequence	Total biomass yield (Mg ha^{-1})	Energy equivalent of economic yield (MJ ha^{-1})	Economic yield (Mg ha^{-1})			Energy input/output ratio based on total economic yield	Benefit/cost ratio
			Sugar	Cereal	Seed cotton		
SD sugarcane/ratoon I/ ratoon II	114.62	1 155 937	254.7	—	—	9.88	2.37
SD sugarcane/ratoon I/ finger millet/cotton	112.15	986 829	195.6	3.83	4.07	8.27	3.13
Finger millet/SD sugarcane/ratoon I/wheat	104.71	961 689	196.5	5.66	—	7.07	2.44
Finger millet/maize/ SD sugarcane/ratoon I	102.71	952 806	185.9	7.73	—	7.86	2.48
ND sugarcane/ratoon I	90.18	901 018	203.6	—	—	10.50	2.47

SD, Short duration variety COA7601 (8-month crop); ND, normal duration variety CO6304 (12-month crop)

many parts of Asia, plant cane and one or two ratoons are followed by the staple crop of rice. However, in Cuba, Hawaii, Guyana, Trinidad, Fiji and Peru, rotation is uncommon since sugar is the sole product. In the energy-starved world of Asia and Africa, sugarcane needs to be rotated with pulse and oilseed crops.

A limited experiment conducted by this author suggests that sugarcane fits well into an agroforestry system. Agroforestry involves multiple cropping with planting of at least one perennial species. Multi-purpose tree species like *Acacia albida*, *A. auriculiformis*, *Albizia falcataria*, *Casuarina equisetifolia*, and *Sesbania* sp. can be grown on the bunds and field borders of sugarcane plantations. *Sesbania rostrata* can be used as a green manure crop in cane fields. Some of the leguminous trees supply ample nitrogen to the cane.

Chapter 5 Nutrient Requirements and Fertilization

Sugarcane makes heavy demands on the replaceable soil nutrients and presages a loss in soil fertility due to monocropping, inadequate fertilization and non-recycling of the crop residue. Barnes (1974) showed that a 50 t crop removes, on average, 30–40 kg N, 22.7–27.2 kg P_2O_5 and 68 kg K_2O. More recently, Zende (1990) has demonstrated the nutrient removal per ton of cane (Table 5.1).

In the past attempts have been made to restore the fertility of sugarcane soils by the use of organic manure, filter cake, bagasse and even oil cake. The use of bulky manures, like farmyard manure, compost, press mud, vinasse (fermentation waste or alcohol slops) and green manuring with sunhemp (*Crotalaria juncea*) *Sesbania aculeata*, *S. rostrata* and *Gliricidia* spp. etc., are important components of sustainable agriculture. More than four decades of research inputs have gone into investigating the nutrition of sugarcane and have facilitated the proper use of organic and inorganic fertilizers, including the time and method of application to improve the yield and quality of cane as well as maintaining soil fertility. Much emphasis is placed on balanced fertilization. In the recent past compound formulations have been increasingly used since they are cheaper on a unit nutrient basis. Aerial spraying of fertilizers and the use of tracer techniques to discover the fate of applied fertilizers are some important milestones in cane culture. Soil and tissue tests are used to guide fertilization in many sugarcane-growing countries. In South Africa the Fertilizer Advisory Service (FAS) has been established to evaluate, monitor and recommend fertilization for the whole cane cycle (Wood 1989).

5.1 Nitrogen

Nitrogen occupies a unique position in plant growth because large amounts are required by most agricultural crops. It is also the primary nutrient influencing the yield and quality of cane. Each year the world's farmers spread roughly 80 million tons of nitrogen fertilizer on to their crops to 'jumpstart' growth. Nitrogen increases the source capacity, namely, leaf area (LA), leaf area duration (LAD), leaf area index (LAI), early canopy closure and the rate of photosynthesis. A higher LA, a function of the number and size of leaves, is primarily responsible for improved biomass production. Increased yield following addition of N is due to a higher tiller number and

Table 5.1. Removal of nutrients from soil during sugarcane growth. (Zende 1990)

	N	P_2O_5	K	Ca	Mg	Na	SO_4
Nutrient removal ($kg\,t^{-1}$ cane)	0.56–1.2	0.38–0.82	1.0–2.5	0.25–0.60	0.20–0.35	0.02–0.2	2.0–2.7

Table 5.2. Effect of N on growth and yield of sugarcane (cv. CO957). (Abayomi 1987)

N level ($kg\,ha^{-1}$)	Leaf area (cm^2)	Tiller number 8 weeks after planting	Stalk length (cm)	Stalk diameter (cm)	Millable cane ha^{-1} (10^3)	Leaf water status (%)	Cane yield ($Mg\,ha^{-1}$)
0	309.67	50.96	124.19	8.19	50.87	81.65	25.84
80	351.57	58.00	131.04	8.30	53.87	85.33	28.09
120	357.93	59.83	144.06	8.25	60.75	87.40	22.04
160	411.25	65.67	138.17	8.07	59.54	88.65	33.78
200	401.62	70.08	130.94	8.23	55.25	90.19	31.40

to yield attributes (Abayomi 1987) like stalk length, stem diameter and number of millable canes. Nitrogen increases succulence as evidenced by a higher leaf water status (Table 5.2).

Symptoms of N deficiency are clearly discernible. Humbert and Martin (1955) indicated that N-deficient canes show uniform yellowing of the leaves, retarded growth, stalks of smaller diameter, premature drying and senescence of old leaves.

Sugarcane absorbs both forms of N, i.e. NH_4 and NO_3, and C_4 plants like sugarcane are more efficient absorbers of available soil N than are C_3 plants (Jones 1985). Under conditions of an adequate water supply, NO_3 is absorbed more than NH_4, possibly due to the greater mobility of NO_3-N. The suggestion is mooted that a constant supply of 40 ppm NO_3-N in the soil (soil log) throughout the crop cycle ensures optimum yield and quality of cane.

5.1.1 Nitrogen Concentration and Uptake

The N concentration varies from 0.2 to 1.6% in the aerial portion of the plant. Clements (1980) observed the highest concentration of 1.77% in the meristem and this declines with age of the plant due to the production of internodes which have the lowest N concentration (0.11 – 0.17%)

Nitrogen uptake is governed by two factors: N demand by the plant, and the supply of available soil N. The availability of soil N is complicated by environmental conditions, genotype and the stage of development. Jones

(1985) showed that C_4 plants can absorb N from a very low concentration (21×10^{-6} *M*) in the soil; in a normal soil the N concentration is greater than 50×10^{-6} *M*. Van Dillewijn (1952) asserted that the maximum amount of N is absorbed within 90 days of planting and N uptake is much ahead of dry matter (DM) production. From the literature it is gleaned that N uptake by a 12-month crop approximates 200 kg ha^{-1}. Assuming a DM production of 60 Mg ha^{-1} with an average N composition of 0.4% and an uptake efficiency of 80%, on a crude basis the N supply works out to 300 kg ha^{-1}. This figure errs on the higher side since roots, trash and other crop residues are invariably returned to the soil.

The augmentative effect of N on DM production or cane yield is apodeictic, but excessive N leads to pithiness and 'piping'. Succulent tissues suffer an increased incidence of pests and diseases with a consequent reduction in juice quality. Ergo, optimum N fertilization ensures a healthy plantation. Stanford and Ayres (1964) envisioned the internal N requirement (INR)[1] of cane; this is defined as the average N uptake (kg) per ton DM or per ton of cane. It is contended that INR is independent of N levels, location and varieties. But for N-sensitive varieties, the INR for near maximum cane yield is higher than that corresponding to sugar yield. The authors concluded that INR averages 1.81 kg per ton DM or 0.91 kg per ton of millable canes.

5.1.2 Nitrogen Use Efficiency

The nitrogen use efficiency (NUE) is defined as the biomass produced or CO_2 fixed per unit of N in the plant. Brown (1978, 1985) made a significant contribution by observing that C_4 plants have a greater NUE than C_3 plants. A reduced investment in storage protein in C_4 species may lead to the higher NUE. Brown (1985) concluded that at very low soil N levels, C_4 species can produce more growth and are more competitive than C_3 species.

NUE is of paramount importance due to escalating costs and the hazards of environmental pollution. It can be estimated by several methods, one of which is the percentage N recovery and is given as:

$$\frac{\text{Nx} - \text{No}}{\text{Fx}} \times 100,$$

where No is the total N in control plot receiving no N fertilizer, and Nx is the total N in a crop receiving Fx amount of N fertilizer.

Another method is to use ^{15}N-enriched fertilizer to calculate the percentage recovery. Novoa and Loomis (1981) argued that the agronomic efficiency (AE) is a better index than NUE.

[1] INR is the uptake of N (kg) per ton of DM produced at mean maximum yield (95–98%). Mathematically, INR is 1/slope of the regression line between the independent variate N uptake and the dependent variate dry matter yield (Stanford and Ayres 1964).

$$AE = \text{Physiological efficiency} \times \text{recovery fraction (RF)}.$$

$$AE = \frac{\text{kg biomass or kg CCS}^2}{\text{kg N applied}} = \left(\frac{\text{kg biomass or kg CCS}}{\text{kg N absorbed}}\right) \times \left(\frac{\text{kg N absorbed}}{\text{kg N applied}}\right).$$

$$RF = \left(\frac{\text{kg N absorbed}}{\text{kg N supplied}}\right) \times \left(\frac{\text{kg N supplied}}{\text{kg N applied}}\right).$$

This implies that AE can be increased by increasing the physiological efficiency, RF or both. RF represents the uptake efficiency, which depends on root properties such as root length, diameter, branching, age of roots, etc, whilst the physiological efficiency is dependent on N efficiency in biomass formation, carbohydrate partitioning, nitrate reduction efficiency and remobilization of protein nitrogen (Novoa and Loomis 1981).

Employing the formula suggested by Novoa and Loomis (1981), the RF for sugarcane (cv. CO62175) ranged from approximately 0.5 to 0.6.

Assuming an average biomass production of 60 Mg ha^{-1} and CCS production of 20 Mg ha^{-1}, an N uptake of 200 kg ha^{-1} and RF of 0.5, the agronomic efficiency for biomass and CCS is 300 and 100 kg per kg N, respectively.

The efficiency for biomass production of 300 kg per kg N absorbed is much higher than the values reported for many crops like wheat, rice and triticale. However, this compares better with an efficiency of 203 kg per kg N in C_4 grasses like Bermuda grass (Brown 1978). NUE can be increased by selection of a suitable genotype, determination of the correct time of application, point placement, reduction of NO_3 losses by using nitrification inhibitors, etc. Emphasis is now placed on nitrogen utilization efficiency (NUTE) rather than on NUE. NUTE is estimated by the formulae given by Trenbath (1986).

$$\begin{aligned} \text{NUTE} &= (\text{N captive efficiency}) \times (\text{N conversion efficiency}) \\ &= (\text{Na/No}) \times (\text{B/Na} \times \text{Hin}), \end{aligned}$$

where No = potentially available N per unit area;
Na = absorbed N per unit area;
B = whole plant biomass;

$$\text{Hin} = \text{nitrogen harvest index} = \frac{\text{N content in stem}}{\text{total N uptake by plant}}$$

[2] CCS, commercial cane sugar.

The experiment conducted by Dinesh Kumar and Hunsigi (unpubl.) in red sandy loam soil with two varieties gave the following results:

N levels (kg ha^{-1})	Potentially available (N_o)	Absorbed N (Na)	Bio-mass yield B	Nitrogen harvest index (Hin)	Nitrogen utilization efficiency (NUTE)
(i) CO62175	⟶	kg m^{-2}	⟵	—	—
150	0.296	0.132	0.496	0.85	1.43
200	0.291	0.175	0.546	0.86	1.61
250	0.249	0.200	0.644	0.86	2.21
(ii) CO7804					
125	0.276	0.166	0.522	0.86	1.63
250	0.268	0.157	0.620	0.87	2.02
375	0.268	0.175	0.654	0.89	2.17

It is gleaned that N fertilization did not significantly increase potentially available N (No) but improved N absorption (Na) by plant and biomass. The nitrogen harvest index (Hin) increased marginally from 0.85 to 0.89 as N increased from 150 to 375 kg N ha^{-1}. The Hin should be kept above 0.5 and NUTE above 2. It is reaffirmed that 250 kg N ha^{-1} is an optimum N dressing for this region.

5.1.3 Interaction with Genotype and Other Nutrients

Varieties show a differential response to applied N and, as indicated earlier, INR does not vary much. Stanford and Ayres (1964) found that the NUE of six cultivars ranged from 526 to 588 kg cane per kg N. But the external N requirement (kg, either cane or sucrose, per kg N applied) varies to a large extent. It is proposed that sugarcane cultivars should be selected under conditions of low soil N availability. Gascho et al. (1986) observed that among four cultivars tested, CP65-357 was most vigorous and far superior in its cane and sugar yield; higher cane yield is attributable to a higher number of millable canes. The external N use efficiency in CP65-357 was 167 and 24 kg for cane and sugar, respectively, per kg of applied N compared with values of 62 and 7 kg for cane and sugar, respectively, in CP63-588. The authors observed that the superior yield of cane and sugar in CP65-357 could be attributed to a high total N accumulation (63 kg ha^{-1}) in the above-ground parts of the plants; this in turn maintained a high plant density until harvest. They reaffirmed that the foliar N concentration should be maintained above 15 g kg^{-1} throughout the grand growth phase.

Srinivasan (1989) proposed that, in general, early varieties had a higher production potential than mid-late or late varieties (Table 5.3). He advocated use of varieties such as CO6806 (early), CO7201 (early) and CO-6304 (mid-late) which have a high production potential at both low and high

Table 5.3. Production of cane ($Mg\,ha^{-1}$) for every additional 100 kg N ha^{-1}. (Srinivasan 1989)

	Change in cane yield ($Mg\,ha^{-1}$) for every additonal kg N ha^{-1}			
Varietal group	First 100 kg	Second 100 kg	Third 100 kg	Fourth 100 kg
Early group (mean of 7 varieties)	33.3	18.5	5.2	– 4.6
Mid-late group (mean of 5 varieties)	35.6	14.7	3.7	– 10.2

input levels of N. On average, these varieties produced 0.5 Mg cane per kg N invested.

Among nutrient interactions, N × K, N × P and N × S are known to influence the yield and quality of cane. Balanced N + K fertilization has been emphasized by Ingram and Hilton (1986) as maximizing the efficient use of inputs. Many studies have revealed that N application in the NH_4 form in conjunction with banded super phosphate has enhanced P uptake. The explanations advanced are: (1) increased solubility of soil P in acid-forming NH_4-based fertilizers; (2) increased root growth in the fertilized zone; and (3) increased metabolic activity of roots. It is also probable that NH_4 application may increase the number of P carriers in plant roots. Additional evidence is given by Thomas and Scott (1990) who observed that N fertilization results in increased P and K uptake. It is likely that N addition stimulated growth of roots and shoots and, ipso facto, increased the area of uptake.

Interaction of N and S in sugarcane has been well detailed by Fox (1976) and the optimum N/S ratio is 10 – 17. Sulphur-deficient plants are typically low in sugars and accumulate NO_3 and soluble forms of other elements with a consequent reduction in recoverable sugar.

5.1.4 Nitrate Reduction

Nitrate taken up by plants needs to be reduced to NH_3 before it can be combined with carbon skeletons to form amino acids (Jones 1985). This reduction is catalysed by two enzymes, namely, nitrate reductase (NR), which reduces NO_3 to NO_2, and nitrite reductase (NiR), which reduces NO_2 to NH_3. Normally, the activity of NiR is several-fold higher to ensure a low concentration of NO_2 which is toxic to plants. In sugarcane, NR activity is lower in roots than in shoots (Maretzki and Dela Cruz 1967). Furthermore, NR activity is higher in young cane stalks than in older ones. Abayomi et al. (1988) found that water stress dramatically reduces NR activity and CO957 has a higher NR activity than CO1001.

5.1.5 Slow-Acting N Fertilizers

Nitrogen losses from sugarcane soils vary from 15 to 27%. In sandy loam soils the loss of N can be over 30% when N is applied at 300 kg ha^{-1}. Hence, the search continues for fertilizers giving reduced losses and improved N recovery in the plant. Tisdale et al. (1985) have categorized the N compounds which become slowly available as follows:

1. Low-solubility compounds requiring decomposition.
2. Sparingly soluble minerals.
3. Soluble substances that gradually decompose.
4. Coated materials.
5. Ion exchange resins.
6. Nitrification and urease inhibitors.

Early workers suggested that the boosting of sugarcane yield by chlorinated hydrocarbons like Telodrin and Gamma BHC was essentially due to inhibition of nitrification (Srivastava 1966). But Parish (1965) had stressed that N serve to be more effective in cane fields by reducing nitrification losses. Among the coated materials, sulphur-coated urea (SCU) has shown considerable promise and the release of urea depends on the thickness of the coatings of elemental sulphur and wax and on the imperfections in the coatings (Wild 1988a). But SCU is relatively expensive. Another important nitrification inhibitor is neem oil obtained from the tropical tree, neem or margosa (*Azadirachta indica*), a native species of India and Myanmar[3]. Hamilton (1992) prophesizes that 'third world entrepreneurs may have a financial winner growing in their backyards and neem tree would boost the economies of the developing countries'. Neem oil contains the alkaloids, *nimbin* and *nimbidin* which inhibit nitrification. *Azadirachtin*, a tetranortri terpenoid, has also been isolated (Stone 1992). Studies in India have amply demonstrated the ability of SCU and neem-coated urea (NCU) to improve the yield and quality of cane as well as leading to enormous savings in the use of N.

Trials conducted in the red, sandy loam soils of Mandya (tropical India) indicated that slow-release N fertilizers like SCU and NCU produced slightly more than 90% of the recommended dose with a saving of 100 kg ha^{-1} (Table 5.4). The slow-release fertilizers were applied in two splits, i.e. in the 6th and 14th weeks after planting, while the recommended practice is four splits, i.e. at planting and in the 6th, 10th and 14th weeks after planting. Urea briquettes proved superior. Juice quality remained unaffected with use of slow-release N fertilizers.

More recent trials suggest that urea super granules (USG) and large urea tablets (LUT) show greater promise in increasing N recovery in sugarcane but this awaits further confirmation (Figs. 5.1–5.4).

---

[3] Neem oil contains limonoids, a class of chemicals that act as deterrents and growth regulators in insects.

Table 5.4. Effect of slow-release N fertilizers on yield and quality of cane (cv. CO62175). (Hunsigi unpubl.)

Treatment	N level ($kg\,ha^{-1}$)	No. of splits	Cane yield ($Mg\,ha^{-1}$)	Percentage of control	Pol in juice (%)
SCU	150	2	139	92	20.79
NCU	150	2	137	91	21.52
Urea briquettes	150	2	161	107	21.49
Urea gypsum (UG)	150	2	127	84	20.48
Control (recommended practice)	250	4	151	100	21.46
F test			NS	NS	
SE			20.86	0.57	

Fig. 5.1. Effect of urea super granules (USG) on sugarcane

5.1.6 Biological Nitrogen Fixation

The pioneering work of J. Dobereiner and her associates in Brazil in the early 1960s clearly established that considerable N fixation takes place in the rhizosphere of sugarcane by non-nodulating bacteria; this is often described as 'associative' or 'non-symbiotic' N fixation. Such associative N fixation is

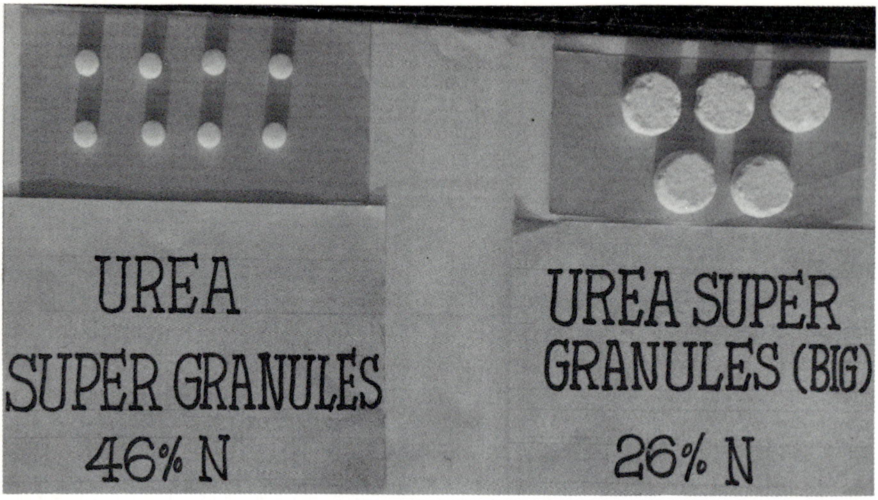

Fig. 5.2. Effect of large urea tablets (LUT) on sugarcane

Fig. 5.3. Effect of neem cake coated urea (NCU) on sugarcane

Fig. 5.4. Effect of urea briquettes on sugarcane

termed 'diazotrophic rhizocoenosis' and results in the addition of 30–75 kg N ha^{-1} yr^{-1} in sugarcane fields. According to Boddey et al. (1991) biological N fixation could be above 200 kg ha^{-1} yr^{-1}. The promising N fixers in sugarcane are *Azospirillum* and *Azotobacter*. Srinivasan (1989) concluded that the late-maturing variety (CO6304) responded to *Azospirillum*, while early and mid-late varieties (CO8001, COC671, CO62174, CO8201) responded to *Azotobacter*.

Varietal differences in N fixation were noticed by Ruschel and Ruschel (1977) and by Abayomi et al. (1988). The latter group observed that CO957 was more responsive to biofertilizers than was CO1001. Lima et al. (1987) showed that CB47-89 was able to obtain large contributions of biologically fixed N.

Based on a large number of field experiments, Srinivasan (1989) has categorized the varieties in respect of their responsiveness to biofertilizers (Table 5.5). Interaction between the soil type and the responsiveness of biofertilizers was reported by Srinivasan and Naidu (1987). In clay soils *Azospirillum* is more effective in N fixation and in loams and sandy soils *Azotobacter* is more effective, thereby improving cane yield and quality.

The main factors which adversely affect N fixation by these bacteria are high O_2 potential, soil moisture stress and low light intensity. Quoting several workers, Jones (1985) concluded that a part of DM increase could be

Table 5.5. Differential response of varieties to biofertilizers. (Srinivasan 1989)

Biofertilizer	Varieties
Responsive to *Azotobacter*	CO419, CO740, CO775, CO62175, CO8143, CO8001, COC671, CO8201, CO62174
Responsive to *Azospirillum*	CO1148, CO7204, CO7704, CO7717, CO8145, CO8209, CO6304
Responsive to both	CO449, CO975, CO62175, CO6304, CO8021, COC671, B37172
Not responsive	CO453, CO1158, CO6806, CO7201, CO8101, CO8150

ascribed to production of growth hormones like indole acetic acid, Cytokinins and gibberellins by these N-fixing bacteria.

The recent research by Boddey et al. (1991) indicated that a new species of N-fixing bacteria, *Acetobacter diazotrophicus* (formerly called *Saccharobacter nitrocaptans*), is most versatile since it has unique physiological properties, such as tolerance to low pH and to high sugar and salt concentrations (osmotolerance). It tolerates short-term exposure to NH_4 and behaves like an endophyte. Hence, this small gram-negative bacterium can be propagated within stem cuttings of sugarcane.

5.1.7 Nitrogen sources, Time and Method of Application

In general, different N carriers have no effect on the yield and quality of cane. A 4-year trial (Table 5.6) conducted in the red soils of Mandya (tropical India) confirmed this statement (Husz 1972).

Additional evidence led Blackburn (1984) to conclude that sugarcane does not show marked differences to different sources of fertilizers except under specific conditions. The anhydrous and aqua forms of ammonia are the cheapest sources but are attended by problems with application. The most commonly used fertilizers are ammonium sulphate (21% N) and prilled urea (46% N) and the usual method of application is by band placement.

In many sugarcane-growing countries ammonium sulphate or urea is applied deep in the furrow at planting and is then covered. Point placement is a better method of application and involves placement of prilled urea in the cane furrows manually in holes made by a wooden peg (1–1.5 m long); this is popularly known as the Java method.

The rationale behind the timing of N fertilization is based on two factors, namely, the small foraging capacity of roots at the initial stages (sett roots), and a major need at the tillering phase. Hence, adequate early fertilization is a key to obtaining high cane and sugar yields. For a 12 to 14-month crop, N should be applied within 60–90 days of planting (Van Dillewijn 1952; Humbert 1968). However, for a 2-year crop cycle the last application was at 11 months and this resulted in a high CCS yield (Borden 1948). When N is applied within 90 days, two or three splits are made depending on the soil

Table 5.6. Relative performance of different N sources on yield of plant cane (cv. CO419). (Husz 1972)

N source	Cane yield ($Mg\,ha^{-1}$)	Average CCS[a] ($Mg\,ha^{-1}$)
Calcium ammonium nitrate	133.0	19.2
Ammonium chloride	134.2	17.7
Urea	130.0	19.0
Ammonium sulphate nitrate	150.0	21.0
Chilean nitrate	136.5	18.0
Ammonium sulphate	127.0	18.5

[a] CCS, commercial cane sugar.

type, with sandy soils receiving three splits while loams and sandy loams receive two splits.

5.1.8 Nitrogen and Juice Quality

Judicious N application is of paramount importance to the production of high sugar yields. A high N dose applied late often results in water shoots (*lalas*) with a high content of reducing sugars and lowered juice purity with a consequent reduction in net rendement (recovery of raw sugar). Clements (1980) has convincingly demonstrated that the N index (N content in 3–6 leaves) and sheath moisture (moisture content of 3–6 leaf sheaths) should drop considerably in order to realize a high sugar recovery. At the grand growth period, adequately manured and well-managed crops should have a sheath moisture of 85% and an N index of 2%. As the ripening process starts, at harvest the sheath moisture should be maintained at 71–72% and the N index should be at 1.0–1.25%.

The ratio of tons of cane to tons of sugar (TC/TS) should be around 5–6 and the primary index (PI), which is the total sugar (reducing sugars and sucrose) content of the young leaf sheaths (3–6) expressed on a dry weight basis, should be 10–11%. In essence, flushing out tissue N and moisture is a prerequisite to increasing sucrose concentration in the stalks.

5.1.9 Response of Sugarcane to Nitrogen

The ubiquity of response of sugarcane to N is well recognised. Attempts have been made since early times to quantify this in dose–response curves. The inverse relation between nitrogen and yield is one such attempt. Capo (1974) has described the optimum fertilizer dosage, based on tissue composition, by employing the Mitscherlich equation.

$$L = A(1 - C^{M}),$$

where L is leaf N content, A is the maximum possible yield and M is the total quantity of nutrient in the soil expressed in fertilizer units.

Capo (1965) further expressed the relation between fertilizer and yield in sugarcane as:

$$Y = \frac{A}{1 + B\,(X - C)^2},$$

where Y is Cane yield, A is the maximum yield and X is the quantity of fertilizer applied. B and C are constants.

The equation is symmetrical and has its maximum value X = C such that C is the level at which the fertilizer element is balanced in respect of other factors. Based on this equation, Capo (loc. cit.) concluded that the N required for maximum cane yield was 1180 lb acre^{-1} (1325 kg ha^{-1}) in Puerto Rico.

A more direct approach to estimating the N needs of a crop is given by Stanford (1973):

$$Nc = Ni + Nm + Nf,$$

where Nc is the uptake of N for the maximum or most profitable yield, Ni is the initial quantity of mineral N, Nm is the estimated N mineralized during the crop season and Nf is the amount of fertilizer N required.

The assumption is that the fractional use, that is, e of all sources of N, is the same:

$$Nc = ei\ Ni + em\ Nm + ef\ Nf.$$

Stanford (1973) showed that:

$$Nf = (Nc/ef) - (Ni + Nm).$$

In the Alfisols of Mandya (India) the N uptake by sugarcane (cv. CO62175) was 204.94 kg ha^{-1}. Ni was 31 ppm or 62 kg ha^{-1} and Nm was 42.4 ppm or 84.8 kg ha^{-1}.

As a first approximation ef was assumed as 0.5:

$$\begin{aligned} Nf &= \left(\frac{Nc}{ef}\right) - (Ni + Nm) \\ &= \left(\frac{204.94}{0.5}\right) - (84.8 + 62.0) \\ &= 263\ \text{kg N ha}^{-1} \text{ for a 12-month crop.} \end{aligned}$$

The exponential or power functions are commonly used to predict the N requirements of cane. The Cobb–Douglas equation is given as:

$$Y = aN^b$$

$$\text{or}\quad \log Y = a + b \log n,$$

where Y is the cane yield in Mg ha^{-1} and n is the N applied in kg ha^{-1}; a and b are constants.

The fitted equation was (Hunsigi 1982a):

$$\log Y = 2.0809 + 0.0284 \log N \; (R^2 = 0.65\,\mathrm{ns}).$$

The fit was poor as evidenced from the non-significant (NS) R^2.

The serious limitation of this equation is that it does not provide the maximum yield and therefore fails to describe the complete range of responses (Pesek 1974). It makes no provision for soil test values.

Following Bray (1965), soil test values were included in the prediction of cane yield with fertilizer addition by the modified Mitscherlich equation:

$$\ln(A - Y) = A[1 - e^{-c_1 b - cx}],$$

where A is maximum yield, b is the soil test value and x represents added fertilizer. c_1 and c are the coefficients associated with the soil test and fertilizer units respectively. The Fitting Mitscherlich's equation (Hunsigi 1982a) to sugarcane (cv. CO419) gave:

$$\log(A - Y) = 147[1 - e^{-0.0253b - 0.0090x}].$$

Bray (1965) stated that relatively immobile ions like P, K, Ca and Mg follow the Mitscherlich–Baule percentage sufficiency concept, while mobile ions like NO_3-N follow Liebig's law of limiting nutrients.

Many empirical models have been fitted in the past to describe the yield response of crops to applied N. Thornley and Johnson (1990) emphatically state that the direct description of the observed data is extremely useful. In one study conducted by Dinesh Kumar and Hunsigi (unpubl.) N dose and yield response were fitted to inverse polynomials:

$$1/y = 1/a + b/a \times n$$

where a and b are constants, y is the yield response of sugarcane and n denotes the N levels.

The fitted inverse polynomial was

$$1/y = 0.00423 + 0.526718\,n.$$

The standard error (SE) for estimated y was ± 0.001224. It is observed that 30% of yield variation can be ascribed to the inverse polynomial. The interpolated values of yield at varying N levels are as follows:

N levels (kg ha^{-1})	125	150	200	250	300	350
Cane yield (Mg ha^{-1})	118.43	129.17	145.09	157.81	167.06	174.37

This author considers that the quadratic model predicts the response of cane to applied N better than does the square root model.

The quadratic or second-degree polynomial equation is given as:

$$Y = a + b_1(x + d) + b_2(x + d)^2,$$

and the square root function is given as:

$$Y = a + b_1(x + d)^{1/2} + b_2(x + d),$$

where Y is the cane yield, x is the fertilizer level and d is the available plant nutrient level.

But d = CT, where C is the fertilizer use efficiency factor and T is the soil test value. a, b_1 and b_2 are constants.

The fitted square root function was:

$$Y = 43.1648 + 8.0954\ (n + d)^{1/2} - 0.1584\ (n + d)$$

$$R = 0.78^{ns}.$$

The rates of N application in the sugarcane-growing regions of the world range from 50 to 300 kg ha^{-1}, and 1 kg N results in an approximate response of 0.5–1.2 Mg ha^{-1}.

5.2 Phosphorus

On a global scale phosphorus is the second most deficient nutrient (Fox 1986) and approximately 35 million tons of phosphatic fertilizers are applied to crops annually. Nearly all tropical and subtropical soils are deficient in P but the response of arable crops to P is unpredictable or uncertain. The luxury consumption of P by sugarcane is a rarity, and the deficiency is manifested by poor tillering and rooting, delayed 'close-in' and shorter internodes which taper rapidly at the growing point. The leaves are narrower with tips that are green-violet or even pinkish or greenish blue due to anthocyanin accumulation (Humbert and Martin 1955). Phosphate compounds in plants are called the 'energy currency' (Tisdale et al. 1988) for they are involved in the formation of energy phosphates. Deficiency of P severely reduces the photosynthetic rate, inorganic P concentration and carbohydrate translocation (Hartt 1972).

Phosphorus is required more by meristematic tissues and the shoot P concentration declines with age. Since much of the P in sugarcane internodes is in a highly mobile inorganic form, the stem P concentration is more sensitive to P nutrition. Hence, Hartt (1959) suggested that the concentration in eight to ten internodes is a better index of the P status of sugarcane.

5.2.1 Fixation of Phosphorus

Plants absorb P in mono-ionic (H_2PO_4) and di-ionic (HPO_4) forms and it reaches the vicinity of the roots mostly by diffusion, although a small fraction moves by mass flow. A high P flux to the roots is possible if the diffusion path is short and there is a large cross-sectional area (Fox 1986). The addition of P through single super phosphate results in several reaction products of low solubility in the soil. Thus, P removal from the solution phase is known as 'fixation'; this collectively describes both sorption and precipitation. Phosphorus sorption depends on several factors, such as the

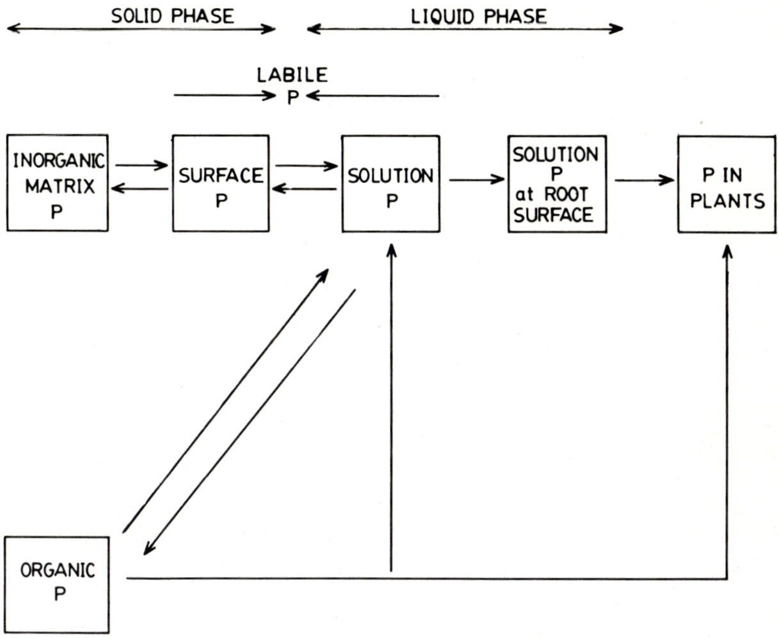

Fig. 5.5. Transfer of phosphate between soil and plants. (Wild 1988a)

type and amount of clay, the cation exchange capacity (CEC), organic matter and content of sesquioxides (oxides of Al and Fe), Ca, Mg, etc. As a rule of thumb, P sorption is associated with a C:P ratio of 300:1, while desorption occurs at a ratio of 200:1 (Wild 1988a). Figure 5.5 presented by Wild (1988a) indicates that inorganic matrix P moves to surface P and then to solution P. The labile pool of surface P, solution P and a portion of organic P may be absorbed by plant roots. Thus, the concentration of P in solution is represented by the intensity factor, whilst sorbed P is represented by the capacity factor.

The external and internal P requirements of sugarcane are high up to 3 – 5 months. The external P requirements is the target P concentration in the soil solution that is associated with the near maximum yield (Fox et al. 1990). Fox et al. (1990) have indicated that the external P requirement of furrow-irrigated cane is 0.005 $mg\,l^{-1}$ and that of drip-irrigated cane is 0.012 $mg\,l^{-1}$. Acceptable sugar yields in the Oxisols of Hawaii can be obtained at the external P requirement of 0.01 $mg\,l^{-1}$ (Fox 1983). The internal P requirement is the percentage of P in the plant tops associated with maximum growth and the values for sugarcane range from 0.16 to 0.19% P.

A simple model for calculating the P requirement of the crop has been proposed by Driessen (1986):

$$D = \frac{Um - Uo}{R},$$

where D is the P fertilizer requirement for sugarcane, Um is P uptake by the cane at an optimum yield level, Uo is P uptake by the cane from unfertilized soil and R is the recovery of applied P fertilizer.

For sugarcane (cv. CO62175) grown in sandy loam soils (Alfisols, Xeralfs) the relevant data are Um = 38.5 kg ha^{-1}, Uo = 17.8 kg ha^{-1} and R = 20% or 0.2, so that:

$$D = \frac{38.5 - 17.8}{0.2} = 103.5 \text{ kg } P_2O_5 \text{ ha}^{-1}.$$

5.2.2 Types of Phosphorus Carrier and Methods of Application

Barring rock phosphate, other phosphatic carriers are equally effective in improving cane and sugar yields in many P-deficient soils. In Mauritius, guanophosphate is used. Several studies have convincingly shown that filter cake (FC) or filter mud (*Dunder* or *Cachaza*) is an excellent P source. FC has the following advantages: (1) it adds organic matter; (2) it leads to better N nutrition; (3) it acts as an ameliorant, increasing the Ca level and decreasing exchangeable Al in acid soils; and (4) it promotes the CEC.

Though FC is highly variable in composition, it is alkaline in reaction. An average of 18 kg per ton of cane with a moisture level of 55–65% can be taken as an approximation. Thus, globally, 17 million tons FC are produced annually with Asia and Oceania accounting for 7 million tons. Prasad (1976) has presented the chemical composition of filter press mud or FC obtained from three countries (Table 5.7).

Prasad (1976) asserted that FC is more effective than triple super phosphate but is no substitute for liming. A level of 20 t FC ha^{-1} improves the yield and quality of cane and is optimum in many soil types. Five tons of FC are assumed to be equivalent to 50 kg soluble P_2O_5. Observations by Ng Kee Kwong and Deville (1988) supported the inference that FC is superior to triple super phosphate as shown in Table 5.8.

Some recent studies have indicated that FC obtained from a sulphitation process is superior to that obtained from a carbonation process.

Three types of P fertilizer, i.e. super phosphate, ammonium phosphate and nitrophosphate, account for nearly 90% of P fertilizers used (Tisdale et al. 1988). In neutral to alkaline soils, water-soluble phosphates are superior to insoluble phosphates. In acid soils, rock phosphate is the best choice. Sett and root treatment have also been attempted. Ammonium phosphate and triple super phosphate can be used as foliar sprays in highly P-retentive soils. Any source of P can be enriched when applied through sugarcane trash.

With regard to time and method of application, P is normally applied in a single dose. The peak P requirement is at the tillering and early growth stages when the 'effective soil volume' (soil permeated with roots) is restricted. Frequently, soluble P fertilizers are band placed near the seed piece or stool to reduce P fixation. But Fox (1986) emphasized that when large quantities of insoluble P (rock phosphate) are applied, it is preferable to

Table 5.7. Chemical composition of filter press mud. (Prasad 1976)

Location	pH	Composition (%)						Composition (ppm)		
		N	P	K	Ca	Mg	S	Mn	Fe	Zn
Trinidad	9.05	1.71	0.45	0.48	1.97	0.70	—	2450	5000	224
South Africa	—	1.69	0.90	0.27	1.84	0.37	0.79	898	—	69
Puerto Rico	—	2.12	1.21	0.57	2.38	2.29	—	1000	7500	—

Table 5.8. Yield ($Mg\,ha^{-1}$) of cane and sugar following use of FC and triple super phosphate. (Ng Kee Kwong and Deville 1988)

	No Nitrogen		120 kg N ha^{-1}	
	Triple super P (250 kg P_2O_5 ha^{-1})	FC (25 t ha^{-1})	Triple super P (250 kg P_2O_5 ha^{-1})	FC (25 t ha^{-1})
Cane yield	98.8	115.5	113.1	121.8
Sugar yield	11.25	12.42	12.54	13.39

incorporate P in the entire soil. Additional evidence from the Oxisols of Brazil led Fox (1986) to assert that residual P is as effective as freshly applied P. The experimental evidence demonstrates that insoluble P, such as rock phosphate, can be mixed with the soil, while soluble forms should be band placed to improve the efficacy of applied P utilization.

5.2.3 Response of Sugarcane to Added Phosphorus

Humbert (1968) demonstrated that P removal is in the range 0.18–0.86 kg/t of cane. Ayres (1959), employing the modified Truog method, estimated 28 kg P_2O_5 ha^{-1} (25 lb $acre^{-1}$) as critical, below which a response to applied P is likely. South African workers (Wood 1989) reported the critical values to be in the range 20–31 ppm (45–75 kg ha^{-1}). Genotypic variation in the response to P is perhaps less important in cane (Humbert 1968).

The response of cane to added P varied from 0.05 to 0.25 Mg ha^{-1} kg^{-1} P_2O_5 in sands, sandy loam soils and red soils, excluding the high P-retentive Vertisols. The dose–response curve to applied P followed a second-degree polynomial (Hunsigi et al. 1974b):

$$Y = 99.1600 + 9.2190P - 1.0650P^2 \quad (R^2 = 0.945^*).$$

where R^2 is the coefficient of determination and the star denotes significance at the 5% level.

Increased cane yield following a P dressing is ascribed to a higher number of tillers per bud and to the height of millable cane. It is generalized that P application to sugarcane in different regions varies from about 45 to 225 kg

P_2O_5 ha^{-1}. In South Africa, for low P-fixing soils a standard dressing of 70 kg P ha^{-1} is given. But in high P-fixing soils 280 kg P ha^{-1} is applied with 70 kg in the furrows and the balance is broadcast. South African workers have devised the phosphorus desorption index (PDI) to supplement the soil test value and the additional dose is justified if the PDI value is below 0.4.

5.2.4 Vesicular–Arbuscular Mycorrhizae (VAM)

Several species of micro-organisms are effective in increasing the availability of native and applied P and these are collectively termed the phosphobacterins. Fox and his associates (1990) contend that sugarcane is least mycorrhizae-dependent. However, species of *Pseudomonas*, *Bacillus* and *Streptomyces* are the major microbes responsible for P solubilization. Among the fungi, *Aspergillus niger* and *A. flavus* are known to solubilize P. Recent studies have shown that the phosphobacterium *Bacillus megatherium* var. *phosphaticum* improved the yield and quality of cane when it was mixed with rock phosphate.

5.2.5 Interactions of Phosphorus

The negative effect of P fertilization on Zn uptake is well documented. In low Zn soils, P fertilization depressed Zn uptake. In many parts of Southeast Asia, where cane is grown after paddy, Zn deficiency is commonly encountered following P addition, thus supporting the Zn–P antagonism. Reduced Zn uptake following P dressing is not merely an external phenomenon of Zn–P precipitation but is physiological in nature. Khan and Zende (1977) noted that the Zn–P interaction originated in the plant roots and affected translocation in the upper plant parts. With regard to P interaction, the acidifying effect of S in neutral or alkaline soils is credited for the prolonged availability of added P. But silicon has an indirect influence on P availability. In leached acid soils with high Fe, Al and Mn contents, the toxic levels of these nutrients is reduced in the presence of Si, thereby improving P availability.

5.3 Potassium

This element is fundamental to sugarcane for the synthesis and translocation of proteins and carbohydrates and for the accumulation of sucrose. In a significant contribution, Hartt (1970) conclusively showed that there is decreased translocation of labelled photosynthate in sugarcane under conditions of K deficiency. Its role in water relations is well recognized and a late K dressing is advocated against severe moisture stress. Potassium application provides a prophylactic measure against eye spot disease caused by

Helminthosporium sacchari. Poor root development in K-deficient plants may be the causal factor of the syndrome known as '*lahaina*' disease. Potassium has also proved effective in correcting the physiological disorder known in Java as *kalamati* disease and which is due to iron toxicity (Von Uexkuell 1968). Lodging in cane is greatly contained by adequate K fertilization. Increased tonnage following K addition parallels sucrose accumulation. Humbert (1968) asserted that an increase in the net rendement following K application is due to a reduction in leaf N (unused N) and to the conversion of reducing sugars to sucrose.

A deficiency of K is seen first in older leaves as it is highly mobile. The leaf margins and tips become brown with necrotic spots which coalesce and show typical 'marginal firing'. In the Philippines increased stalk mortality is attributed to K deficiency.

Sugarcane removes enormous quantities of potash from the soil, upto about 900 kg K_2O ha^{-1} (Humbert 1968), although this includes luxury consumption. On average a ton of millable cane removes 0.4–2.5 kg K (Von Uexkuell 1968). Unlike uptake of N and P, K uptake by the plant is slow in the early stages but reaches a maximum between 3 and 7 months.

Supply of K to plant roots is by diffusion and nearly 90% reaches the roots in this way. But diffusive K flux is reduced at low water contents (high PF) and O_2 potentials and at high clay contents (Grimme 1976). Root extension and proliferation are equally important since K is absorbed by an active process.

Potash in the soil is held at three different exchange sites or binding sites (Fig. 5.6). Figure 5.6 shows that K in planar sites (P-position), edge sites

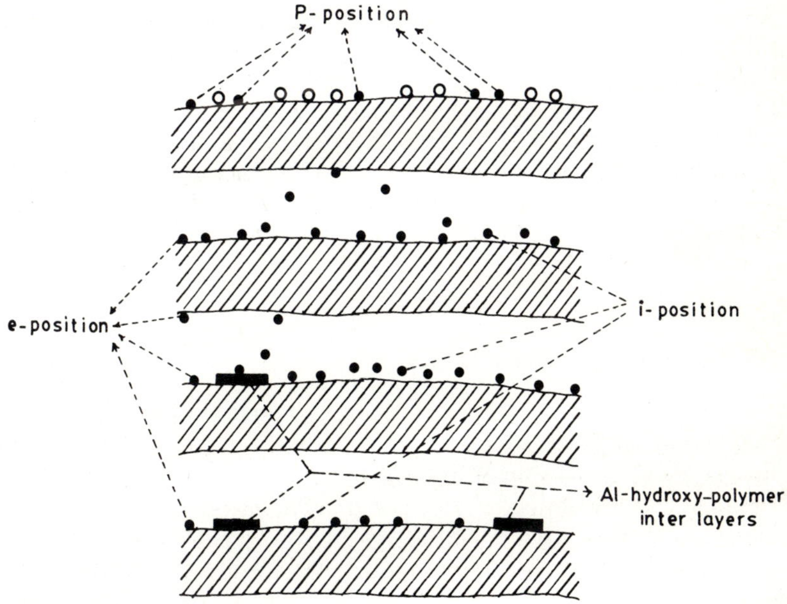

Fig. 5.6. Three different exchange sites for potassium

(e-position) and inner binding sites (i-position) has low, medium and high K selectivity, respectively. As a consequence, K in the P-position, and to some extent K in the e-position, is in equilibrium with solution K, whereas K in the i-position is tightly held and constitutes the structural or matrix K.

5.3.1 Forms of Soil Potassium

Soil K is categorized into four different forms in terms of availability to plants (Fig. 5.7). These forms are matrix or structural, non-exchangeable (fixed or difficult to obtain), exchangeable and solution K, and they exist in equilibrium shown in Fig. 5.7. An equilibrium is quickly established between solution and exchangeable K, but equilibrium between the exchangeable and non-exchangeable phases is rather slow. Addition of soluble K fertilizers increases the solution K. Since sugarcane is known to devour soil K, solution K is rapidly depleted and gradually exchangeable and non-exchangeable forms get into the solution phase causing a net soil K depletion. Nevertheless, the K balance in the soil can be maintained by returning crop residues like sugarcane leaves, sheaths and trash which contain a sufficient quantity of K.

5.3.2 Quantity/Intensity (Q/I) and Quantity/Potential (Q/P) Isotherms in Sugarcane Soils

The Q/I buffer curve stems as a corollary from Schofield's ratio law and in simple terms may be defined as the ratio of K activity to the activities of Ca

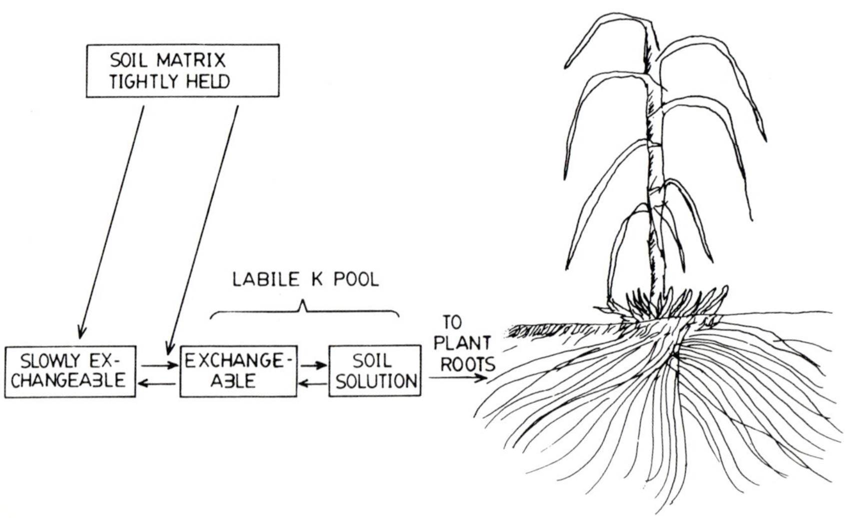

Fig. 5.7. Different forms of available potassium in the soil

and Mg on an equivalent basis in a non-saline soil. It is a good measure of K availability to plants. The activity ratio, AR_0K, is expressed as:

$$AR_oK = aK/a\,(Ca + Mg)^{1/2},$$

where aK and a (Ca + Mg) are the activities of K and Ca plus Mg.

The advantage of the Q/I approach seems to be in the simultaneous measurement of the quantity (capacity) factor and the intensity factor (rate of release). A characteristic Q/I buffer curve is depicted in Fig. 5.8. The parameters involved in the curve are:

ΔK – Quantity by which the soil gains or loses K in reaching equilibrium (quantity factor, Q).

$\Delta_0 K$ – Linear part of the curve which, when extrapolated, meets the K axis, i.e. K held in the p-position and readily exchanged.

Kx – K held in specific potassium sites.

KL – labile K pool.

PBCK – Potential buffering capacity of K (slope of Q/I curve).

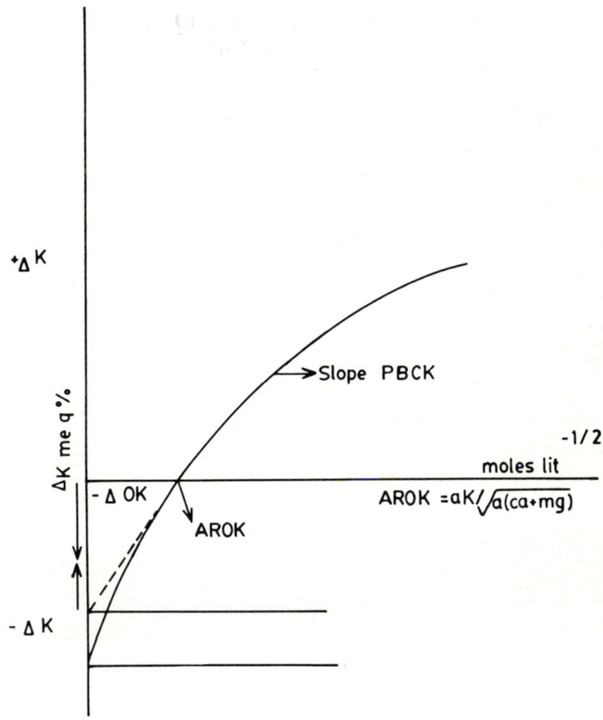

Fig. 5.8. Typical Q/I buffer curve for potassium

In fact KL correlates most commonly with the exchangeable K, and PBCK reflects the non-exchangeable K reserves in the soil and is well correlated with the clay content (Beckett 1971).

Studies conducted by Hunsigi and Srivastava (1980) have revealed that in some sugarcane soils AR_0K (intensity factor) varied from 0.65×10^{-3} mol $l^{-1/2}$ in Sankeshwar soil to 2.1×10^{-3} mol $l^{-1/2}$ in IISR farm soil (Table 5.9). A wide range of AR_0K values, i.e. 0.1×10^{-3} to 12×10^{-3} mol $l^{-1/2}$, have been reported by Fernandez Caldas et al. (1975) for volcanic ash soils.

Another important Q/I parameter is the PBCK value which measures the resistance to change in AR_0K (i.e.buffering) and is related to the concept of interconnected wells.

Table 5.9 shows that PBCK ranges from 54 to 191 mEq/100 g (mol $l^{-1/2}$) in five soils. Fernandez Caldas et al. (1975) reported very high values of 468mEq/100 g (mol $l^{-1/2}$). In our studies a significant correlation was observed between PBCK and CEC (r = 0.979). The soils of Sankeshwar and Gangavati (both Vertisols) had a higher CEC and hence possessed a high PBCK. The labile K pool (KL) serves as a 'vehicle' by which K from interlattice sites is made available to plants. This author reiterates that KL is a good measure of exchangeable K but is a poor index of K availability to a long-duration crop like sugarcane.

Potash exhaustion study involving continuous cropping with sugarcane (cv. CO1148) was conducted in pot culture by Hunsigi and Srivastava (1980). It was found that the leaves developed typical K-deficiency symptoms such as necrotic spots from the tips downwards with inverted V-shaped 'marginal firing'. The K index (3–6 leaf sheaths) at this stage was less than 0.1% on a dry weight basis. It is suggested that these soils may have reached a critical K concentration (activity), i.e. 0.65×10^{-3} to 0.94×10^{-3} (mol $l^{-1/2}$), at which sugarcane was unable to absorb K. As a first approximation, it is mooted that the critical cation activity ratio (CCAR) for K in sugarcane would be 0.65×10^{-3} to 0.94×10^{-3} (mol $l^{-1/2}$). Thus, the field response of cane to K dressing at these CCAR values can be obtained, and conforms to the observations of Beckett (1972) who produced a comprehensive list of CCAR for several crops, including the CCAR for K in several grasses with values ranging from 0.1×10^{-3} to 0.6×10^{-3} (mol $l^{-1/2}$). Arnold et al. (1968) found that the CCAR for potato was 5×10^{-3} (mol $l^{-1/2}$), which is nearly six times more than the value for sugarcane. It is inferred from this that potatoes need to be fertilized with a higher K concentration in the soil than is necessary for sugarcane.

In order to circumvent the limitations of the soil test method and to provide a universal application, the concept of chemical potential (ΔGK) in the soil was devised. The molar free energy change for K, or ΔGK, is defined as the free energy change associated with replacing one equivalent of K with one equivalent of Ca plus Mg. This can be expressed as:

$$\Delta GK = -RT \ln \frac{aK}{\sqrt{a(Ca + Mg)}},$$

Table 5.9. Some physicochemical characteristics and Q/I and Q/P parameters of some sugarcane soils. (Hunsigi and Srivatsava 1980)

Parameter	Mandya Alfisols (Xeralfs)	Sankeshwar (Vertisols)	Gangavati (Vertisols) natric horizon	Pusa Aridisols (Calciorthids)	IISR farm soils (Aridisols)
CEC (mEq/100 g soil)	11.96	77.22	61.78	9.85	11.72
pH (1:2.5)	6.55	7.85	8.15	7.85	7.25
EC at 25°C ($S\ m^{-1}$)	0.149	0.245	1.112	0.234	0.106
Exch. K N NH_4OAc (mEq/100 g)	0.333	1.026	0.974	0.320	0.487
Exch. K conc. H_2SO_4 (mEq/100 g)	0.308	0.679	0.679	0.179	0.551
Isotopically exchangeable ^{42}K (mEq/100 g)	1.7436	0.2134	0.2422	1.6890	2.4511
K saturation (%)	2.788	1.328	1.577	3.254	4.157
Basal diffraction spacing (d) in Å	6.92	14.25	13.33	10.09	9.66
Dominant clay mineral fraction	Kaolinite	Smectite	Mont-morillonite	Micaillite or chlorite	Illite
Activity ratio, AR_oK ($mol\,l^{-1/2}$) $\times 10^{-3}$	0.94	0.65	0.80	1.35	2.10
PBCK (mEq/100 g; $mol\,l^{-1/2}$)	58.06	183.33	190.91	54.54	57.14
KL (mEq/100 g)	0.2067	0.1097	0.1792	0.0775	0.1282
Chemical potential of K, ΔGK ($cal\ mol^{-1}$)	– 3740	– 3920	– 3400	– 3200	– 3280
Isotopic ratios with ^{42}K (counts of soil solution/counts of standard solution)	0.85	0.23	0.21	0.69	0.96
Absolute velocity constants (KP) in NaCl + NaTPB (h^{-1})	0.0112	0.0252	0.0451	0.0643	0.1046
Dry weight of cane shoots (g/pot) (cv. CO1148)	83.70	88.34	90.08	75.30	87.29
K uptake or K yield (g/pot)	0.6632	1.1030	1.4510	1.3072	1.3263
Field behaviour of sugarcane to K	Both plant and ratoon canes responsive	Both plant and ratoon canes highly responsive	Responsive	Non-responsive	Non-responsive

where aK and a(Ca + Mg) are the activities of K and the common ionic species of Ca plus Mg respectively, T is absolute temperature and R is the gas constant. The initial chemical potential (ΔGK) of some sugarcane soils (Table 5.9) was shown to range from -3920 cal mol^{-1} in the Vertisols of Sankeshwar to -3200 cal mol^{-1} in the Aridisols (calciorthids) of Pusa. Woodruff (1955) classified soils according to chemical potential, with ΔGK of -3500 to -4000 cal equiv^{-1} being associated with K deficiency. But values of -2000 cal equiv^{-1} were associated with excessive K. perusal of Table 5.9 indicates that the Alfisols of Mandya respond to K fertilization, as do the Vertisols of Sankeshwar but the response of Gangavati soils is doubtful. The Aridisols of Pusa and the IISR farm soils do not respond to added K. This corroborates with the field behaviour of sugarcane to K application. It is worth noting that the Aridisols of Pusa and IISR farm soils have K-bearing minerals, i.e. illite (Table 5.9).

Crop plants have well-defined growth stages and at each stage the plant has a work potential which indicates its strength to utilize and assimilate the nutrients in soil (Talibudeen 1974). Thus, the work potential is the measure of the amount of energy expended by the plant in drawing from the soil a nutrient at that potential.

It is surmised that the chemical potential of nutrients in the soil and the work potential of the crop need to be integrated so that fertilization is based on these parameters.

As indicated earlier, the exhaustion potential, or CCAR, for sugarcane was estimated from two soil types, namely the Vertisols of Sankeshwar and the Alfisols of Mandya (Hunsigi and Srivastava 1980, 1983), by continuous cropping with sugarcane (cv. CO1148); values ranged from -4129 ± 78 to -4347 ± 86 cal mol^{-1} (Table 5.9). The different uptake potentials of sugarcane are presented in Table 5.10.

The ΔG optimum of $-1610 \pm$ a SE of 224 cal mol^{-1} for sugarcane is much higher than the values so far reported (i.e. -2500 cal mol^{-1}) for many other crops. Addiscott (1970) pointed out that the ΔG response for potatoes was -4710 cal equiv^{-1} as against -5600 cal equiv^{-1} for rye grass. This means that potatoes need a much higher K concentration in the soil solution than does rye grass for maximum expression of yield potential. The value of ΔG optimum (-1610 cal mol^{-1}) for sugarcane in the sand culture study works out to a K concentration of 5.01 mEq l^{-1}. This is in agreement with values given by Samuels and Alexander (1968) who recommended a K concentration of 5.0 mEq l^{-1} for sugarcane grown in standard nutrient solution.

Table 5.10. Different work potentials of sugarcane (cv. CO1148). (Hunsigi and Srivastava 1983)

	ΔG exhaustion	ΔG optimum	ΔG luxury	ΔG toxic
cal mol^{-1}	-4347 to -4129	-1610	-1093	-946
cal equiv^{-1}	-4589 to -4371	-1851	-1335	-1188

5.3.3 Soil K Extractants to Predict K Supply to Sugarcane

A large array of chemicals, from hot water (Soxhlet extraction) to buffer solutions, alkalis and acids, have been employed to measure K supply to sugarcane. Other methods include percolation techniques and use of exchange resins. Nemeth (1975) observed a good correlation between the yield of perennial rye grass and the amount of K dissolved, by using electro-ultrafiltration (EUF) which measures the intensity factor and is a better measure than the activity ratio (AR_0K). A critical review of soil extractants to predict K availability to sugarcane is presented by Hunsigi and Srivastava (1976).

Several workers, including Ayres and Hagihara (1955), du Toit (1959) and Innes (1959), observed that the exchangeable K obtained from N NH_4OAc (pH 7) is a better index of K availability to sugarcane, and a critical limit was placed at 75–100 ppm. Duflo (1976) contended that in Africa cane grown in soils with less than 120–150 ppm will respond to K dressing. Jones (1985) concluded that other parameters like extractable K in subsoil, CEC, organic matter level, etc. should be considered when making fertilizer recommendations based on soil tests.

Ammonium acetate (NH_4OAc) is universally used as a soil K extractant since NH_4^+ has nearly the same ionic radius (0.143 nm) as that of K^+ (0.133 nm) and the former can become fixed in the hexagonal holes of clay crystals. Hence, NH_4OAc will release only the exchangeable K which, in the opinion of this author, does not satisfactorily predict K availability to a long-duration crop like sugarcane. A portion of non-exchangeable K needs to be included to predict K availability to the plant or ratoon cane (Hunsigi and Srivastava 1981).

A distinction should be made between readily released K (exchangeable K) and that which is slowly released from the non-exchangeable portion. A part of the non-exchangeable K which is available to plants is called 'step K', while constant rate K (CRK) reflects the structural or matrix K. The specific soil K extractant, sodium tetra-phenyl boron (NaTPB), and successive boiling in 1N HNO_3 separate out that portion of K which is available to the plant (step K). It should be borne in mind that NaTPB is a 'clean' chemical, introduces less analytical errors and predicts satisfactorily K availability to sugarcane but it is costly. Successive extractions with boiling 1N HNO_3 introduces large analytical errors. It is therefore apparent that exchangeable K per se is not a satisfactory index of K availability to sugarcane, but a portion of the non-exchangeable K obtained with NaTPB or boiling 1N HNO_3 should be included to improve the prediction of K availability. Wild (1988b) reaffirmed that the prediction of K availability is improved by the inclusion of readily released non-exchangeable K. For routine soil-K testing concentrated H_2SO_4 is a better extractant than NH_4OAc when sugarcane is taken as a test crop. A simplified nomogram (Fig. 5.9) presents the relationship between several systems for measuring the K status in sugarcane soils (Hunsigi and Srivastava 1981).

The relative contribution of exchangeable and non-exchangeable K towards K uptake or K yield by sugarcane (cv. CO1148) grown in different

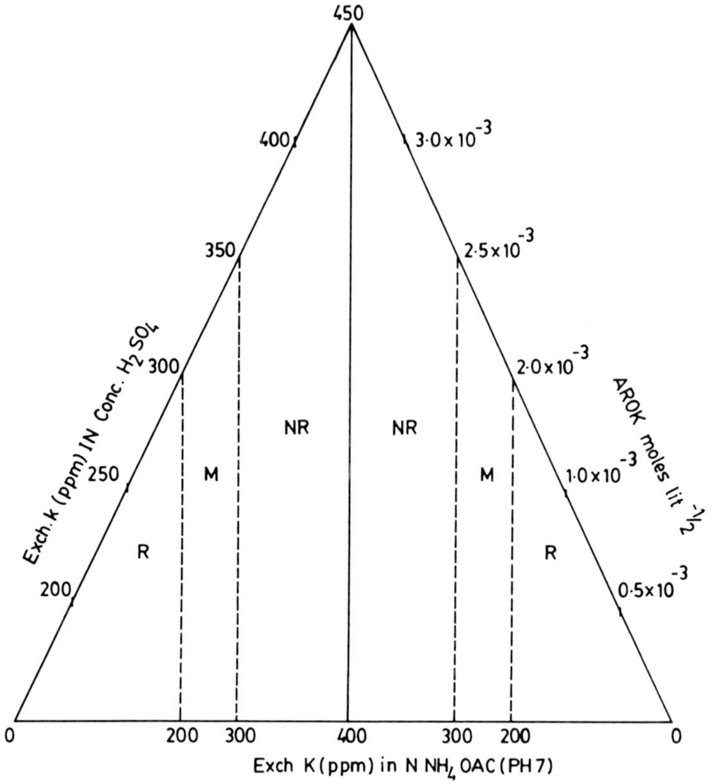

Fig. 5.9. Nomogram to predict K availability to sugarcane. R, responsive; M, marginal; NR, non-responsive

soils under exhaustive cropping is shown in Table 5.11. It is seen that the maximum contribution of exchangeable K was from the Vertisols of Sankeshwar and Gangavathi, whilst the maximum contribution of non-exchangeable K was from the Aridisols of Pusa and the IISR farm soils. In the Alfisols of Mandya, both exchangeable and non-exchangeable K contributed equally towards K uptake by cane.

5.3.4 Field Response of Sugarcane to Applied K

Increased tonnage following K application is reported by Samuels and Landrau Jr. in Puerto Rico (1956), du Toit (1959) in South Africa, Innes (1959) and Innes and Chinloy (1951) in Jamaica, Levirington et al. (1962) in Australia, Tabayoyong and Robinol (1962) in the Philippines, Hodnett (1956) in the West Indian islands, Aloma et al. (1974) in Cuba and Hunsigi et al. (1974a) in tropical India. In South Africa Stewart (1969) observed a response of about 5.0 and 2.4 Mg ha^{-1} of cane and sugar, respectively, at an application rate of 93 kg K ha^{-1}.

Table 5.11. Contribution of exchangeable (Ke) and non-exchangeable (Kne) K in K uptake by sugarcane grown in different soils (exhaustive cropping with sugarcane cv. CO1148). (Hunsigi unpubl.)

Soil type	K yield or K uptake (mg/pot)	Contribution of Ke (mg/pot)	Contribution of Kne (mg/pot)	Contribution of Kne (%)	Contribution of Ke (%)	Initial Ke (mg/pot)
Mandya (Alfisols)	663.2	382.8	280.4	57.7	42.3	780.15
Sankeshwar (Vertisols)	1103.0	1111.0	—	100.0	0.0	2399.90
Gangavati (Vertisols)	1451.0	1278.8	172.2	88.1	11.9	2279.86
Pusa (Aridisols)	1307.2	56.9	1250.3	4.4	95.6	749.97
IISR farm soils (Aridisols)	1326.3	162.8	1163.4	12.3	87.7	1140.05

The response to potash obeys Mitscherlich's relationship and the constant was 0.01 to 0.012 for cane and 0.008 for sugar (Hodnett 1956; Innes 1959). Locsin et al. (1956) used a modified Mitscherlich equation and the second-degree polynomial characterized the potash–yield relationship. du Toit (1968) demonstrated that the responses of cane depended on the available K_2O content of the soil as shown in Table 5.12.

Hunsigi and his associates (1974a) showed that cane (cv. CO740) grown in the Vertisols of Sankeshwar responded to K application. The fitted parabola was:

$$Y = 134.2403 + 9.5296\,K - 1.4199\,K^2 \quad (R^2 = 0.90^{**}).$$

where Y is the expected yield and R^2 is the coefficient of determination; ** indicates significance at the 1% level of probability. In the cultivators' holdings in the same region, a linear response was observed:

$$Y = 83.3699 + 0.1021\,K.$$

However, K responses are unlikely in alluvial soils which contain K-bearing minerals like illite, mica and chlorite.

The agronomic value of K rests in the improvement in girth of cane. In some cases it has improved the millable cane population. With regard to time and method of application, a single band application is preferred. In very sandy soils two splits may be necessary. All K sources are equally effective and our studies have indicated that potassium scheonite (24% K_2O) is as good as muriate or sulphate of potash.

According to Humbert (1962), better juice quality was associated with increased cane and sugar yields following K fertilization and he recommended use of a fertilizer approach involving 2:1:2 of $N:P_2O_5:K_2O$. The optimum K index is kept at 2.25%. Since all the K in cane is dissolved in the cell sap, the K-H_2O index (i.e. K content of 3–6 leaf sheaths expressed as a percentage of tissue moisture) is suggested (Clements 1980). The K-H_2O index measures K content as a percentage of tissue moisture rather than dry weight. This approach reduces the effect of genotype on tissue K concentration and improves K diagnosis (op. cit.).

Interaction of N and K is often encountered under field conditions and adequate K fertilization is essential for improved cane and sugar yield. Potassium improves the sucrose percentage in cane and an improvement in cane quality is possible only when K application results in a drop in leaf N

Table 5.12. Effect of K_2O content of soil on the response of sugarcane

Amount of K_2O applied ($kg\,ha^{-1}$)	Response ($Mg\,ha^{-1}$)	
	Soils < 150 ppm K_2O	Soils > 150 ppm K_2O
112	14.33	− 0.99
224	21.49	+ 1.73

and sheath moisture. Thus, adequate K is required to utilize the assimilated N in cane to bring a stage of maturity when reducing sugars are converted to recoverable sugars. This led Clements (1980) to argue that inadequate K nutrition results in the accumulation of simple sugars and amino acids with an eventual drop in net rendement.[4]

5.4 Soil Test, Foliar Analysis and DRIS to Guide Fertilization

Much research has gone into quantifying the fertilizer requirements of sugarcane and mid-term correction ('fine tuning') has been used to achieve the desired objectives. Soil tests have been used in-extenso in the sugar industry for a long time; a number of soil extractants have been used to monitor fertilization and critical limits have been established.

It is now realized that soil tests and tissue analysis are complementary and help in determining a proper fertilization schedule. Accordingly, critical nutrient concentrations and the choice of tissue sampled assume importance. Clements (1980) is credited with providing a comprehensive system popularly known as crop logging. He defines it in simple terms as a 'crop log, like a ship log is a record of progress from its start until arrival at harvest'. The material used for N determination is 3–6 leaf blades, while for other nutrients 3–6 leaf sheaths are used. Sheath moisture (in 3–6 leaf sheaths) seems to be of prime importance for it integrates all the physiological processes of cane. Another important parameter is primary index (PI). The limits set by Clements (op. cit.) at an early stage of growth (6–10 months) are given in Table 5.13.

It is essential that the sheath moisture and N index drop at harvest to 72–75 and 1.7–1.8%, respectively, to ensure sugar accumulation. Clements

Table 5.13. Critical limits of some nutrient indexes (%) and sheath moisture index (%) at early growth stage. (Clements 1980)

PI (3–6 sheaths)	N index (3–6 leaves)	P index (3–6 leaf sheaths)	K index (3–6 leaf sheaths)	Moisture index (3–6 leaf sheaths)
10.0	2.0	0.08	Above 2.25	85.0

[4] Net rendement is a measure of the amount of refined sugar produced from raw sugar and is evaluated by estimating the pol, ash and reducing sugars. The formula for estimation is empirical and varies from country to country. However, a generalized formula is:

$$\text{Net rendement} = \text{pol} - (3 \times \text{ash}) - (2 \times \text{reducing sugar}).$$

In this text rendement or net rendement simply means sugar recovery.

Table 5.14. NPK requirements of BO varieties based on DRIS values. (Hunsigi 1985)

Variety	Cane yield ($Mg\,ha^{-1}$)	DRIS index N	DRIS index P	DRIS index K	Nutrient requirement	NBI
BO70	36	+ 1.72	− 3.24	+ 1.52	P > K > N	6.48
BO90	36	− 1.77	+ 5.55	− 3.78	K > N > P	11.10
BO91	56	+ 5.2	− 18.98	+ 13.78	P > N > K	37.96
BO99	53	− 0.25	− 5.58	+ 5.38	P > N > K	11.66
lsd 0.05[a] varieties	8.2					
lsd 0.05 DRIS		3.63				

[a] lsd 0.05, least significant difference at the 5% level of probability.

(1980) stressed that the standard N index (SNI) is calculated for standard moisture levels and crop ages for a given variety. An alternative approach is stalk logging, since stalks seem to be more sensitive to tissue concentrations.

More recently, critical nutrient concentrations and soil test methods have been used as a means to determine the crop fertilization requirements. Hence, the diagnosis recommendation integration system (DRIS) has been widely tested in South Africa, Brazil, Florida and Hawaii (Meyer 1981). Elwali and Gascho (1984) obtained an increased yield of cane and sugar in Florida by adopting a fertilization procedure through DRIS.

DRIS was developed by Beaufils (1972) and tested by several workers. The method uses ratios of tissue nutrient concentrations to calculate DRIS indices in order to diagnose nutrient deficiencies and imbalances. The efficiency of DRIS lies in its insensitivity to factors like tissue age, position of leaf, cultivars, etc. It is especially suited to reveal the antagonism among nutrients.

The DRIS indices may be positive or negative but their sum is always zero because they measure the balance among nutrients. The most negative index indicates the nutrient most required and the most positive index indicates the nutrient least required. An index close to zero suggests that the nutrient concerned is present in adequate supply.

The DRIS indices of BO varieties grown in Aridisols (Calciorthids) in subtropical India are presented in Table 5.14. In the last column of this table the nutrient balance index (NBI) denotes the balance among nutrients in each variety. NBI is obtained by adding the values of DRIS indices irrespective of their sign. The larger the value of NBI, the greater is the imbalance among nutrients.

The Fertilizer Advisory Service (FAS) in South Africa has successfully employed the DRIS approach in guiding fertilization of cane and they are investigating DRIS soil norms (Meyer 1981).

The general NPK fertilizer schedules adopted for sugarcane grown in different countries is given in Table 5.15.

Table 5.15 NPK fertilizer schedules used for sugarcane grown in different countries

Country	N ($kg\,ha^{-1}$)	P_2O_5 ($kg\,ha^{-1}$)	K_2O ($kg\,ha^{-1}$)
Argentina	60–120	40–60	—
Australia	112	67–90	78–168
Barbados	75–100	—	—
Brazil	60–120	75	60–80
Cuba	67–134	50–65	70–85
Dominican Republic	34–68 (280 irrigated)	—	—
Egypt	119–262	—	—
Guyana	56–101	60–80	60–80
Hawaii	224–336	225	224–450
India: subtropical	160–200	80–100	0–60
tropical	250–300	100–120	60–120
Indonesia	80–120	—	—
Jamaica	45–168	—	—
Java	130–150	—	—
Martinique	200–250	120–140	140–160
Mauritius	100	—	—
Mexico	120–240	60–120	75–85
Pakistan	135–202	90–135	78–112
Peru	200–300	60–80	40–50
Philippines	90–200	80–200	80–160
Puerto Rico	135–200	50–80	112–120
South Africa	112–224	70 (280 high P-fixing soils)	40–45
Sudan	200	—	—
Taiwan	160–225	75	75
Trinidad	50–150	80–150	150
Venezuela	150–200	—	—

Note: This is a very rough guide and represents average applications. Data collected from several sources.

5.5 Calcium and Magnesium

The essentiality of Ca^{2+} in higher plants was recognized by True (1922), although Clements (1980) stated that Bamford was credited with having indicated its essentiality to plants. The main functions of Ca are: maintenance of structural and functional integrity of plant cell membranes, formation of Ca pectate, counteracting the toxic effects of other ions, etc. A ton of cane removes 0.5–1.0 kg Ca. This element is more of a soil conditioner and has a valuable influence in raising soil pH and improving microbial activity which results in better soil structure. A deficiency in arable neutral soils does not occur but is not uncommon in Oxisols, Ultisols, Andepts and Histisols.

Hence, a ratio of Ca/total cations from 0.15–0.2 is suggested as adequate for meeting the Ca requirement of sugarcane.

The visual symptoms of Ca deficiency appear on the leaves as minute chlorotic spots with dead centres which turn dark reddish brown (Humbert and Martin 1955). Calcium has low mobility in plants. Growth is retarded by Ca deficiency. The rind becomes soft. The stalks are of smaller diameter and taper rapidly at the growing point. The roots of Ca-deficient plants are severely discoloured (Evans 1959).

The critical exchangeable Ca level in soil is 100 ppm and the Ca index (determined for 3–6 leaf sheaths) is 0.14–0.175% (op. cit). But an adequate Ca level is associated with 0.06% Ca in 8–10 internodes (Humbert 1968). The top visible dew lap (TVD) lamina is more sensitive to Ca deficiency and the threshold values are 0.05–0.1%, while a value of 0.18–0.2% TVD in leaf indicates adequate Ca nutrition.

True Ca deficiency in a normal arable soil is a rarity; however, the beneficial effects of liming in acid soils is recorded time and again. Liming corrects acidity, reduces the toxicity of Fe and Mn and improves the availability of P and Mo with a consequent improvement in cane and sugar yields. In acid soils application of lime stone ($CaCO_3$) ranges from 2–3 t ha^{-1}. Other sources of Ca are gypsum ($CaSO_4$) and $CaCl_2$. Excessive liming is accompanied by reduced uptake of Mg and K due to antagonism. Lime induces iron chlorosis. Clements (1980) emphasized that freckling (necrotic leaves) is due to high concentrations of Fe, Mn, Zn and Al salts and of some other metals in sugarcane grown in acid soils which lack Ca.

Magnesium is absorbed as Mg ions and is transported by mass flow. It may also reach the plants by root interception. A ton of cane removes approximately 0.8 kg Mg. A shortage of Mg is accompanied by reduced chlorophyll content and the cane stalks are smaller in diameter with shortened internodes and internal browning (Evans 1959). Magnesium is relatively mobile in plants; hence, spotting is more severe in older leaves and produces a rusty appearance.

Soils are deficient when they contain less than 20–50 ppm exchangeable Mg. But the Ca/Mg ratio is important and should not exceed a value of 7. Hence, Mg deficiency is more probable in highly limed soils or those dressed with K. Evans (1959) observed that in Mg-deficient cane, the TVD lamina contained 0.03–0.05% Mg, but an Mg content of 0.1% in 3–6 leaf sheaths or of 0.05% in 8–10 internodes suggests adequate Mg nutrition.

Widespread Mg deficiency is noticed in the Fujian province of China. The soils are subject to moderate weathering with high erosion and rapid leaching (Qu-Ming and Ju-Ming 1986). The red and lateritic soils contain 0.38 mEq exchangeable Mg/100 g soil, while yellow soils contain 0.47–0.57 mEq/100 g soil. The red soils are of economic importance for many commercial crops are grown and the threshold value seems to be 60 ppm Mg. But the ratio of exchangeable Mg/CEC is a better indicator and a ratio of less than 10% is accompanied by higher responses to applied Mg (Qu-Ming and Ju-Ming 1986). Biswas (1991) observed that many soils in south China are deficient in Mg and more than 20 crops responded to Mg dressing. Sugar and starch crops like sugarcane, sugarbeet and cassava have a high Mg requirement. The sugarcane yield increased from 52.67 Mg ha^{-1} in NPK-treated plots to 102.8 Mg ha^{-1} with application of dolomite, and such practice should be encouraged.

5.6 Silicon

Paddy and sugarcane are highly siliciferous plants of economic importance, and Vandervorm (1980) found that the monocots of the family Graminae are Si accumulators. Silicon accumulation is found in the endodermis of sett and shoot roots of *Saccharum* and allied genera like *Erianthus* and *Miscanthus* (Wynnparry and Kelso 1977). Absorption of Si by roots is in the form of silicic acid [$Si(OH)_4$] and this is an active process. The metabolic exclusion of Si has also been reported. Redeployment of Si in the plant is perhaps unlikely and silicate sprays are ineffective (Fox et al. 1969).

The importance of Si to plants has been recognized since the days of Liebig but its essentiality is a matter of conjecture. The concensus is that it is not an essential element but that it is beneficial to all crops (Wild 1988b). The beneficial effect is due to its structural role and to providing the plant with resistance to insects and diseases. Silicon increases the strength and rigidity of cell walls and leads to reduced lodging. It is also associated with better water relations and increases water-use efficiency. In a significant contribution, Alexander (1973) contended that Si application (in lower concentration) inhibits invertase activity and lowers the level of acid phosphatase, thus improving sucrose synthesis. Wong You Cheong and Halais (1976) observed that the relatively large application of silicate sources, i.e. basic slag, cement, calcium metasilicate, etc., to highly acidic, low silicon-containing humid Latosols had an ameliorative effect on the soil. The multiple benefits of Si addition are:

1. Si applied as calcium metasilicate [$Ca(SiO_3)_2$] raises the pH by neutralizing acidity.
2. P nutrition is improved due to reduced P fixation.
3. There is reduced toxicity of Fe, Al, Mn, Zn and other metals.
4. 'Freckling' is controlled as Si acts as a corrective to excess Fe, Mn, etc. in the plant.
5. Decreased acidity causes better microbial activity and release of organically bound N, P and S.
6. For the well-being of sugarcane, the ratio of Mn (ppm)/SiO_2 (%) in 3–6 leaf sheaths should be kept in the range 30–50.
7. There is improved yield and sugar recovery.

A significant response of sugarcane to Si application has been reported in Australia (Haysom and Chapman 1975), Florida (Elawad et al. 1982a, b), Hawaii (Clements 1965; Ayres 1966; Fox et al. 1969), Mauritius (Wong You Cheong and Halais 1976) and Puerto Rico (Samuels 1969a).

In some areas of Florida and Hawaii, sugar yields correlate well with the Si concentration of index tissues (i.e. 3–6 leaf sheaths) (Bowen 1981; Elawad

Table 5.16. Effect of silicate levels on the yield and quality of cane (cv. CP63–588) in a muck soil of Florida. (Elawad et al. 1982a; values taken only from plant crop)

Rate of silicate application ($t\,ha^{-1}$)	Plant height (cm)	Stem diameter (cm)	Millable stalks ($\times 10^3\,ha^{-1}$)	Cane yield ($Mg\,ha^{-1}$)	Sugar yield ($Mg\,ha^{-1}$)
0	112	2.6	89	93	9.2
5	128	2.8	107	126	12.3
10	131	2.9	114	144	14.0
15	132	3.0	121	156	15.6
20	135	3.1	131	167	16.6

et al. 1982a). Calcium metasilicate at 4–5 t ha^{-1} increased both cane and sugar yields. These benefits are associated with an increase in photosynthetic area, tillering and millable canes and with better P nutrition and decreased Al toxicity. Ayres (1966) argued that there is a threshold for the soluble Si content in soils ($\sim$166 ppm) for better expression of yield potential in cane. However, according to Wong You Cheong and Halais (1976), yield correlates less to soil silicon than to leaf sheath silicon and the threshold value is 3.6% SiO_2 of dry matter. The observations of Fox et al. (1969) confirm that the sheath Si index is more sensitive than that of the leaves or internodes.

The response of silicate-containing amendments can be anticipated in highly weathered, leached tropical soils (Oxisols and Ultisols) with low values of pH, CEC, extractable Si and base saturation, and in organic soils (Histisols). Elawad et al. (1982b) demonstrated that there were no significant differences in silicate sources like TVA slag, Florida slag and Portland cement. Silicate material at a concentration of 15 t ha^{-1}, broadcasted and thoroughly mixed in the soil, increased the cane and sugar yields by 68 and 79%, respectively in a plant crop (Table 5.16). Increased yield following silicate application is attributable to increases in plant size, plant height, stem diameter and the number of millable stalks.

A multi-year response of sugarcane to calcium silicate slag in the Histisols of the Everglades (Florida) was reported by Anderson et al. (1991). They observed that 20 t slag ha^{-1} increased the cumulative cane yield by 39% and the sugar yield by 50% over three crop years.

5.7 Sulphur

This nutrient is taken up almost entirely in the SO_4 form by an active process. It reaches the roots by mass flow and to a smaller extent by root interception. Both in the developed and developing world, S was an 'incidental' constituent of many fertilizers. But in the recent past it has been found that prolonged use of S-free fertilizers led to S deficiency. Tisdale et al. (1985) record that soils in several African countries, like Nigeria, Senegal and

Upper Volta, are deficient in S. Fox (1976) observed S deficiency in Puerto Rico, Zimbabwe, Hawaii and Louisiana.

Sulphur exists in the soil in three forms, namely organic, soil solution and oxides of Fe and Al (allophane). An available source of S to crops is as SO_4 in soil solution. This is replenished by desorption from soil clay and hydrous oxides. It is logical to assume that sulphate adsorption is beneficial for it prevents or minimizes losses from leaching. Organic sulphur is mineralized and this depends on the temperature, moisture, pH, etc. Mineralization is associated with a C:S ratio of 200:1 (N:S ratio of 10:1 or 20:1). However, immobilization takes place at a C:S ratio of 400:1. Recent evidence shows that 50% of organic S in the surface soil is present as organic sulphate esters (R–O . SO_3). Sulphatase enzymes hydrolyse these esters and release S in the SO_4 form (Tisdale et al. 1985):

$$R\text{–}O.SO_3 + H_2O \xrightarrow{\text{Sulphatase}} R.OH + H_2SO_4.$$

(organic sulphate ester)

Sulphur from the atmosphere has become increasingly important. Normally rain adds 10 kg S $ha^{-1}yr^{-1}$. Due to atmospheric pollution, S compounds are present both in the gaseous form (SO_2) and in the particulate or aerosol form. Industrial areas add more S to soil than rural areas.

Sulphur is associated with chlorophyll formation (Dutt 1962) and a deficiency affects carbohydrate metabolism. According to Ghosh et al. (1990), S is directly connected with N utilization since it can improve N-use efficiency, possibly by increasing nitrate reductase activity. Symptoms of S deficiency are similar to those of N deficiency, save the older leaves remain green in contrast to the yellowing of younger leaves. This suggests a lower mobility of S than N. A typical symptom of S deficiency in leaves is general chlorosis with green midribs. Sulphur deficient plants possess narrow, pointed leaves and shorter, thinner stalks with reduced tillering. Such plants show anthocyanin pigmentation.

The requirement of cane is about 0.91 kg per ton millable cane or 1.81 kg per ton DM and varietal differences have been noted (Humbert 1968). The varieties H44-3098 and H38-2915 showed less sensitivity to an S shortage than did H49-5; variety H37-1933 was most sensitive to S deficiency.

Stanford and Jordan (1966) state that uptake of S by sugarcane parallels DM production. They argue that the S requirement for growth is well reflected by a normal ratio of organic N to organic S. Hence, a good indication of S deficiency is a decline in the N:S ratio of tissues to below 17:1 because the ratio of protein N:protein S is 15:1.

In a more systematic study, Fox (1976) determined the internal and external (SO_4-S) S requirements for sugarcane as 40 and 10 kg ha^{-1}, respectively. A strong, positive interaction between N and S is obvious. Leaf sheaths (3–6) provide a better index of S nutrition than do leaf blades. It is generalized that the critical S concentration in sugarcane ranges from 0.5 to 0.8% in 3–6 leaf sheaths or from 0.08 to 0.15% in 3–6 leaf blades.

Table 5.17. Effect on cane and sugar yield of application of sulphur to the soil. (Anon. 1987a,b)

S levels ($kg\,ha^{-1}$)	Cane yield ($Mg\,ha^{-1}$)	Sugar yield ($Mg\,ha^{-1}$)
0	31	3.0
25	87	11.2
50	123	15.5
100	148	18.3

Tandon (1991) has proposed a critical value for available soil S as 20 ppm for sugarcane and responses can be anticipated below this threshold value. In Australia, improved cane and sugar yields were observed following S application (Table 5.17). Application of 25 kg S ha^{-1} yr^{-1} is recommended by Chapman (1985) and other Australian workers (Anon. 1987a, b). Increased yield following S application is due to improved root weight per cane and a higher number of tillers per bud. However, juice quality remained unaffected.

Fox (1976) made the practical suggestion that the seed pieces from fields well supplied with S would perhaps meet the S requirement of this crop. Wherever a severe S deficiency is observed, 20–30 kg elemental sulphur may be mixed in to the soil prior to planting.

5.8 Micronutrients

Based on field experience, nutrient deficiencies are generalized with respect to sugarcane nutrition in Table 5.18.

It is perhaps surprising to note that Na is essential for C_4 plants, including sugarcane, for it increases the activity of phosphoenolpyruvate carboxylase (Tisdale et al. 1985). Chlorine has a principal role in water relations and Bowen (1972) observed that the Cl level in sugarcane is quite high and is recognized as a major element.

Micronutrients reach the plant roots mainly by mass flow and diffusion (Fig. 5.10). Deficiency symptoms for Zn, Cu, Mo and B have been developed

Table 5.18. Probable nutrient deficiencies in sugarcane

Deficiency	Nutrient
Very common	N
Common	P, K, Fe, Zn
Occasional	Mg, S, Mn
Rare	Ca, Cu
Unknown	Mo, Cl, Na, B

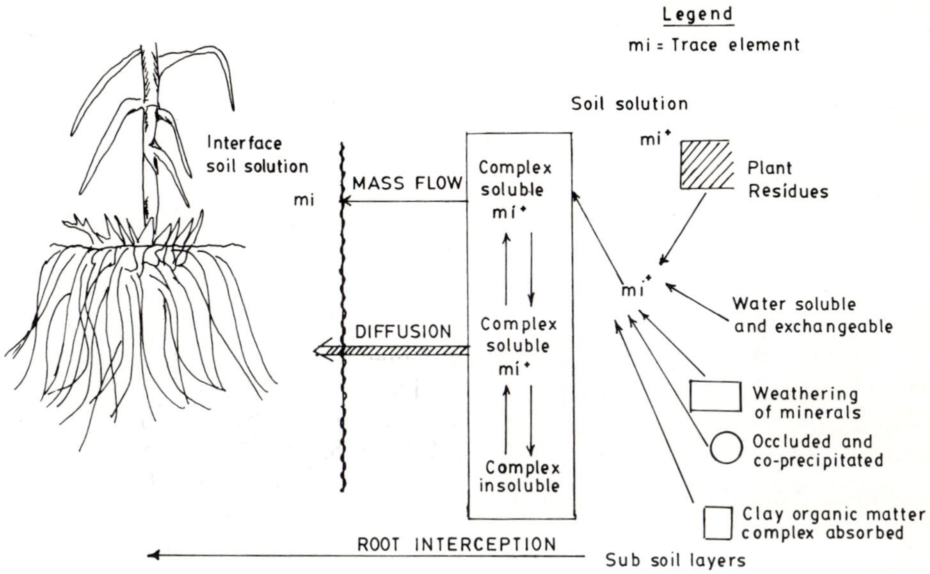

Fig. 5.10. Movement of trace elements in soil to reach the vicinity of the roots

in a sand/water culture by Agarwala et al. (1986, 1987). They observed a shortening of the internodes and chlorosis of the leaves due to a deficiency of Zn, Cu and Mo with depressed activity of catalase. A deficiency of Mo resulted in maximal reduction of catalase activity. Boron deficiency caused a marked reduction in sucrose in cane stalks. However, under field conditions, a deficiency of Fe, Mn and Zn is normally encountered in sugarcane.

5.8.1 Iron

In general, monocots are more susceptible to Fe deficiency than dicots, and Fe functions both as a structural component (in chloroplasts) and as a co-factor for enzymatic reactions. This author believes that a true Fe deficiency in cane is quite rare since the sett itself contains sufficient iron to support early growth. However, lime, carbonate/bicarbonate, P and heavy metal induced Fe chlorosis is frequently met in sugarcane estates. In Hawaii, Mn-induced Fe chlorosis is seen, whilst in the Histisols of the Everglades in Florida Cu-induced Fe chlorosis is seen since copper sulphate has been applied since 1927 (Barnes 1974). Other factors which restrict Fe availability are high pH (7.4–8.5), excessive soil moisture and poor aeration.

The visual symptoms of Fe deficiency are a general paling of young leaves followed by alternating green and chlorotic stripes. Chlorosis following Fe deficiency is due to inadequate chlorophyll synthesis and is associated with haem compounds (Yonachen and Barak 1982). Furthermore, iron–manganese interactions are well recognized. In the red soils of Fiji or the Latosols of Hawaii, normal cane has an Fe:Mn ratio of 15:1 or more. But Fe deficiency or Mn toxicity results in a ratio of 1:1 or even less (Evans 1959). Our observations indicate that Fe-deficient plants accumulate Fe in the nodal region and mobility in these tissues is very much restricted. Iron chlorosis can be ameliorated by soil application of $FeSO_4$ (25–50 kg ha^{-1}) or by use of foliar spray (2% $FeSO_4$). Iron chelates like FeEDTA, FeEDDHA and NaFeDTPA are more effective than iron salts. But Saroha and Singh (1980) demonstrated that elemental S is far superior to all other sources. The physiological explanation offered by these authors is that under conditions of induced chlorosis, Fe does not remain in a physiologically available form in the tissues. Hence, applied S reduces the leaf sap pH by increasing the S content and/or decreasing the Ca content and makes Fe more mobile in the tissues. Band placement of S is better than broadcasting and 500 kg S ha^{-1} has improved the yield and quality of cane (Saroha and Singh 1980). These authors concluded that sulphur application 'created a balanced nutritional environment and forestalled the chemical or biological inactivation of Fe' and thus improved the yield and quality of cane.

5.8.2 Zinc

Sugarcane grown after paddy rice in many parts of Asia has shown Zn deficiency, which is manifested by stunted and patchy growth. Intensive production practices coupled with use of complex fertilizers lacking in trace elements have lead to zinc deficiency. The functional roles attributed to Zn are auxin metabolism and involvement in many enzyme systems. The distinct deficiency symptoms are a deterioration in the spindle, young leaves tend to furl and the stalks become severely 'piped' with a light brown colour (Clements 1980). Severe Zn deficiency results in a drastic reduction in cane and sugar yields. Liming induces Zn deficiency as Zinc solubility decreases with a rise in pH. Antagonism of P on Zn uptake is well documented. Apart from the low solubility of zinc phosphate, the ability of plants to regulate P accumulation is either lost or impaired when there is a zinc deficiency. Hence, excess P accumulation (some times at toxic levels) causes symptoms similar to those of Zn deficiency (Tisdale et al. 1985). High concentrations of Fe, Mn and especially Cu cause severe Zn deficiency. This can easily be overcome by application to the soil once every 2–3 years of commercial-grade zinc sulphate at 20–25 kg ha^{-1}. Even a 1% $ZnSO_4$ foliar spray is recommended for sugarcane at the 3-month stage.

The improved yield and quality of cane following soil or foliar application of $ZnSO_4$ was noted by Kumaresan et al. (1985) under field conditions. For an average of seven locations the response was around 4 Mg ha^{-1} and a foliar spray was superior to soil application. Among sugarcane varieties, CO6304 responded more, in terms of increased tonnage, to micronutrient application than did COC671 and COC8201.

5.8.3 Manganese

This differs from other micronutrients in that both a deficiency and toxicity can be seen under field conditions. Acid soils possess Mn at toxic levels, while liming causes a deficiency and the borderline between the two is thin. Interveinal chlorosis is a distinct symptom of Mn deficiency in cane. Such a deficiency also causes 'freckles' or long, uniform necrotic lesions surrounded by dark areas (Clements 1980).

In a detailed investigation Bowen (1983) established critical nutrient concentrations (Table 5.19) in 3–6 leaf sheaths and associated the values with tons cane ha^{-1} (TCH) and tons sugar ha^{-1} (TSH). He reiterated that the Si concentration in sheaths per se is of no consequence in determining the cane or sugar yield. But the ratio of Mn/SiO_2 is important in determining the yield and quality of cane with the optimum being 50–75. Bowen (1983) observed that Mn/SiO_2 has a profound influence on the ratio tons cane/tons sugar (TC/TS). At a Mn/SiO_2 ratio of 25, the ratio of TC/TS is 10.11 but as the Mn/SiO_2 ratio increased to 150, so the TC/TS ratio dropped to 8.57. This led Bowen to conclude that cane tonnage decreased with an increasing ratio of Mn/SiO_2 although the cane had a higher sugar content. It may further be noted that the critical levels suggested by Bowen do not vary much from those reported by Clements (Table 5.19).

5.8.4 Copper

A deficiency is rare occurring infrequently on calcareous or overlimed soils. Copper deficiency in sugarcane is well recognized by a 'droopy top' and the unfurling of spindle leaves. In Cu-deficient cane, the stalks are soft and rubbery.

5.8.5 Boron

The prime role of B in plants is in the transport of sugars across the cell membranes (Sugar-borate complexes) and thus it helps to improve cane quality. This author has observed an increased pol in juice following use of boron sprays (1 kg B ha^{-1}) on cane grown in sandy soil, although the effect was not statistically significant.

5.8.6 Molybdenum

A deficiency of Mo in field-grown sugarcane is almost unknown. However, Alexander (1973) proved that Mo treatment increased the sucrose content by suppressing the activity of ATPase and glucose-6-phosphatase.

Table 5.19. Critical levels of macro-and micronutrients in elongating leaf sheaths (3–6) associated with maximum yields of cane and sugar. (Bowen 1983; Clements 1980)

Nutrient[a]	Critical level in 3–6 leaf sheaths		
	Clements (1980)	Bowen (1983) TCH	TSH
Ca (% dry wt)	0.17	0.2–0.25	0.20
Mg (% dry wt)	0.08	0.10	0.10
S (% dry wt)	0.22	0.30	0.30
B ($\mu g\,g^{-1}$ dry wt)	2.0	2.0	2.0
Zn ($\mu g\,g^{-1}$ dry wt)	10.0	10.0	10.0
Cu ($\mu g\,g^{-1}$ dry wt)	10.0	5.0	8.0
Mn ($\mu g\,g^{-1}$ dry wt)	10.0	10.0	10.0
SiO_2 (% dry wt)	—	2.2	—
Mn/SiO_2 (Mn $\mu g\,g^{-1}$ dry wt/ SiO_2 % dry wt)	—	—	50

[a] The figures are on a soluble sugar-free basis.

5.9 Integrated Nutrient Management

Modern agriculture, not withstanding sugarcane culture, is at a crossroads of two extreme scenarios. One way is to resort to organic farming or natural farming, where chemical fertilizers are totally dispensed with, while the other way is to use high doses of fertilizers which pollute the air, water and land. The middle way, namely, integrated nutrient management (INM), seems to be the panacea to achieve the desired objectives of high yield and quality of crops with sustainable productivity. This author is firmly convinced that with organic farming the nutrient supply is not adequate to meet the demands of sugarcane within 100–120 days. Even with the best organic farming system a reduction in yield of 20–25% is anticipated. The other alternative is bio-intensive farming which aims to use bio-fertilizers and a deep tillage practice to ensure maximum root proliferation for exploitation of subsoil nutrients and water. Varieties need to be dove-tailed with expansive root systems for efficient utilization of soil nutrients. But INM, a happy blend between bulky manures, chemical fertilizers and bio-intensive farming, is a better choice.

Optimum soil moisture and improved physical conditions have a profound influence on nutrient uptake and utilization. The optimum bulk density of 1.2–1.3 $kg\,m^{-3}$ and an O_2 potential of 10–20% in the soil increase the uptake of many nutrients and thus ensure their efficient use.

Balanced fertilization is an integral part of INM. Bajwa and Qidwai (1986) have shown that balanced fertilization is a prerequisite for INM as shown in Table 5.20.

This implies no deficiency or antagonistic effect of nutrients but rather a synergistic effect. It is postulated that fertilization based on DRIS values will

Table 5.20. The role of balanced fertilization in INM. (After Bajwa and Qidwai 1986)

N	P_2O_5 (kg ha^{-1})	K_2O	MgO	Cane yield[a] (Mg ha^{-1})	Yield increase (%)
170	0	0	0	15.5	—
170	114	0	0	27.1	74
170	114	114	0	34.1	120
170	114	114	25	39.7	160

[a] Cane yield is averaged over 3 years.

improve fertilizer use efficiency (FUE) in sugarcane. Since sugar is the ultimate product, early and adequate fertilization with N, P and other nutrients at peak and critical times are a prerequisite. Experience has shown that application of bulky manures, like compost and farmyard manure, at the rate of 20–25 t ha^{-1} have increased the nutrient supply of major, minor and secondary elements in many tropical and subtropical soils. Application of press mud helps not only to reclaim alkaline soils but also supplies adequate P to cane. Investigations by this author demonstrated that the incorporation of leguminous intercrops, like sunhemp, French beans, green gram, etc. reduced the requirement of fertilizer N to the extent of 50–100 kg ha^{-1}, besides improving soil fertility.

Chapter 6 Irrigation and Drainage

Sugarcane is classed among those plants having a high water requirement and yet it is drought tolerant. The duty of water for cane is fixed at a minimum of 200 acres ($\approx$ 80 ha). It grows well in areas receiving solar energy from 18 to 36 MJ m^{-2}. Rough estimates show that 80% of water loss is associated with solar energy, 14% with wind and 6% with temperature and humidity. The literature is replete with water requirement studies and values vary from $\sim$ 1000 to $\sim$ 3000 mm (Table 6.1) depending on the agro-ecological conditions and crop cycle (12–36 months). The amounts of water required to produce 1 g cane, DM and sugar are 50–60, 135–150 and 1000–2000 g, respectively. The peak water use of 0.75–0.85 cm day^{-1} is observed during the grand growth phase. In Hawaii, Chang et al. (1965) recorded daily levels of evapotranspiration (ET) of 0.3–0.5 cm in winter and 0.7–0.9 cm in summer. A linear relation exists between water use and DM production, and between ET and yield/sugar (Shih 1985).

Certain physiological characteristics need to be reckoned:

1. A prodigious water supply reduces the yield of cane and/or sugar while mild water stress enhances the yield (Turner 1990).
2. Excessive watering at tillering should be avoided since it coincides with active root development and hinders nutrient uptake due to poor O_2 diffusion.
3. A 'drying off' or 'cut-out' period for 4–6 weeks prior to harvest ensures an optimum sugar yield. Abstraction of water during the 'ripeness to flower' stage helps to control tassels.
4. Sugarcane has an intrinsic ability to circumvent water shortages and can more than make up for the potential yield loss. The term *rattrapage* is advanced for the ability of cane to offset an initial setback. However, at later stages, an adequate water supply is necessary to ensure complete *rattrapage* and a full cane yield (Govinden and Arnason 1990).
5. When cane is intercropped with potatoes, groundnut, tomatoes, beans or soyabeans, an additional 20–40% water is required.

6.1 The Root System

Initially the sett and shoot roots are superficial. Later, the sett roots become freely branched and act as superficial absorbing roots, and finally the buttress and rope systems of roots are formed. Moisture extraction is from

Table 6.1. ET of sugarcane grown in different areas of the world

Region/area	(mm ET yr^{-1})	Reference
South Natal	1267	Thompson (1976)
Pongola, South Africa	1555	
Mauritius (under drip irrigation)	450–500	Hodnett et al. (1990)
Hawaii	2000–2400	Chang et al. (1965)
Australia	1522	Thompson (1976)
Cuba	1681–2133 (plant cane) 1193–1679 (ratoon cane)	Fonseca and Egana (1989)
Taiwan	1500–2200	Chang et al. (1968)
Morocco	1000	El Messaoudi (1990)
Puerto Rico	1752	
Jamaica	1387	A review presented by Thompson (1965)
Venezuela	2420	Ewart (1965)
Philippines	2451 (52 weeks)	Parthasarthy (1972)
Subtropical India	1400–1800	Srivastava and Johari (1979)
Tropical India	2000–2400	
Burundi, Central Africa	1327–2017	Hunsigi (unpubl.)

Original values were maximum rates of potential ET in in. day^{-1} and have been converted to mm yr^{-1}.

soil to a depth of 60 cm. The growth rate usually declines with an increase in the age of the root system. Hudson (1968) even suggested that there is a critical root density to meet the transpirational need of 0.6 cm day^{-1}. Varietal differences exist with regard to root proliferation and extension.

Mongelard (1968) demonstrated that Ebene 1/37 has a less developed root system than M147/44. Similarly, COC671 and CO997 have a less developed root system in sandy loam soils than does CO62175. This study further indicated that under furrow irrigation nearly 90% of soil moisture extraction takes place in the first 60-cm of the soil layer (Fig. 6.1). Under drip irrigation the roots are spread on the surface and the cane is more prone to lodging. A greater percentage of roots in deeper soil layers are found in unirrigated plots than in irrigated plots (Batchelor et al. 1990). It is reasonable to assume that lighter soils tend to encourage a dense and deep root system, whilst heavy soils promote less dense and more superficial rooting.

A valuable contribution was made by Thompson (1976) who distinguished between total available soil moisture (TAM) of 10 kPa to 1.5 MPa and freely available moisture (FAM) of 10 kPa to 0.1 MPa. The elongation rate of sugarcane roots should be maintained at an optimum of 40 mm day^{-1} in lighter soils to utilize the receding soil moisture.

As a general rule the available soil water for different textured soils is:

heavy textured soil = 200 mm per m depth of soil
medium textured soil = 140 mm per m depth of soil
coarse textured soil = 60 mm per m depth of soil

From the additional experiments we have conned that modified Penman is the most reliable method to estimate ET of the crop. In Nigeria it was

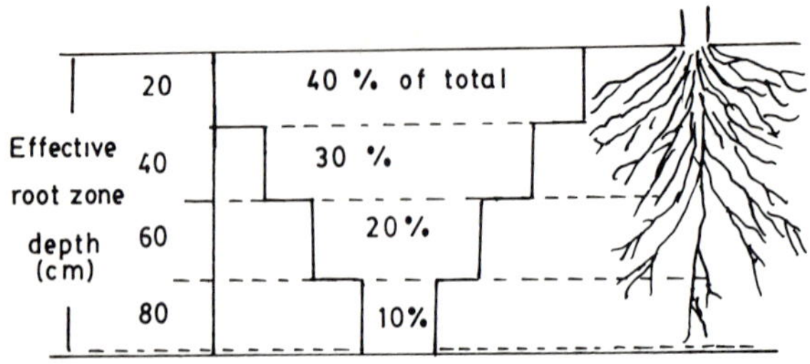

Fig. 6.1. Moisture extraction pattern of sugarcane (cv. CO62175) grown in a sandy loam soil

found that the ET (Penman) ranged from 4.52 mm d^{-1} in December to 6.85 mm d^{-1} in April. The maximum evapotranspiration (ETm) is the maximum ET when the crop is fully supplied with water and adequate nutrition and calculated as:

$$ETm = K_c \cdot ET \text{ (Penman)}$$

Where K_c is the crop coefficient which increases from a minimum value of 0.5 at crop emergence to a maximum value of 1.15 at full canopy and declines to 0.6 at maturity. Based on the above equation the water requirement for 12 month crop in Bacita Estate (Nigeria) worked out to 1770 mm (Jaiswal, S. P., Pers. comm). Further, the actual evapotranspiration (ETa) equals ETm when available soil water (ASm) to the crop is optimum. However ETa < ETm when ASM is limited.

6.2 Scheduling Irrigation

The earliest method for scheduling irrigation to cane was Das (1933) day-degree (DD) concept and DD was the number between daily maximum temperature and 70 °F. Sugarcane was irrigated when the accumulated DDs reached 350. Historically, resistance blocks of gypsum, nylon, fibre glass units, Bouyoucos meters and neutron probes were employed on sugar estates to schedule irrigation with varying degrees of success. In heavier soils irrigation was resumed when the resistance block readings were 5000 Ω (3.699 log Ω) but the maximum yield of cane or sugar was achieved when irrigation commenced at 2700 Ω (3.4 log Ω). An array of tensiometers have been successfully used in sugarcane fields and irrigated when the average readings were −25 to −35 kPa.

A climatological approach caught the imagination of many agronomists after the potential evapotranspiration (PET) concepts of Penman (1948) and Thornthwaite (1948) were developed. A significant correlation was observed between the PET values of Penman and Thornthwaite and the tank evaporation (Table 6.2 and Fig. 6.2) in the Mosso region of Burundi, Central Africa (Hunsigi, unpubl.).

Table 6.2. Correlation between tank evaporation and PET values from different formulae in the Mosso region of Burundi, Central Africa (Hunsigi unpubl.)

Empirical method	Symbol for PET	Total ET for sugarcane (cm yr^{-1})	Correlation coefficient, r	F test
Penman (1948)	e	133.5	0.92	Sig $\propto$ 0.01
Thornthwaite (1948)	p	71.97	0.58	Sig $\propto$ 0.05
Blaney and Criddle (1950)	Cu	159.11	0.37	NS

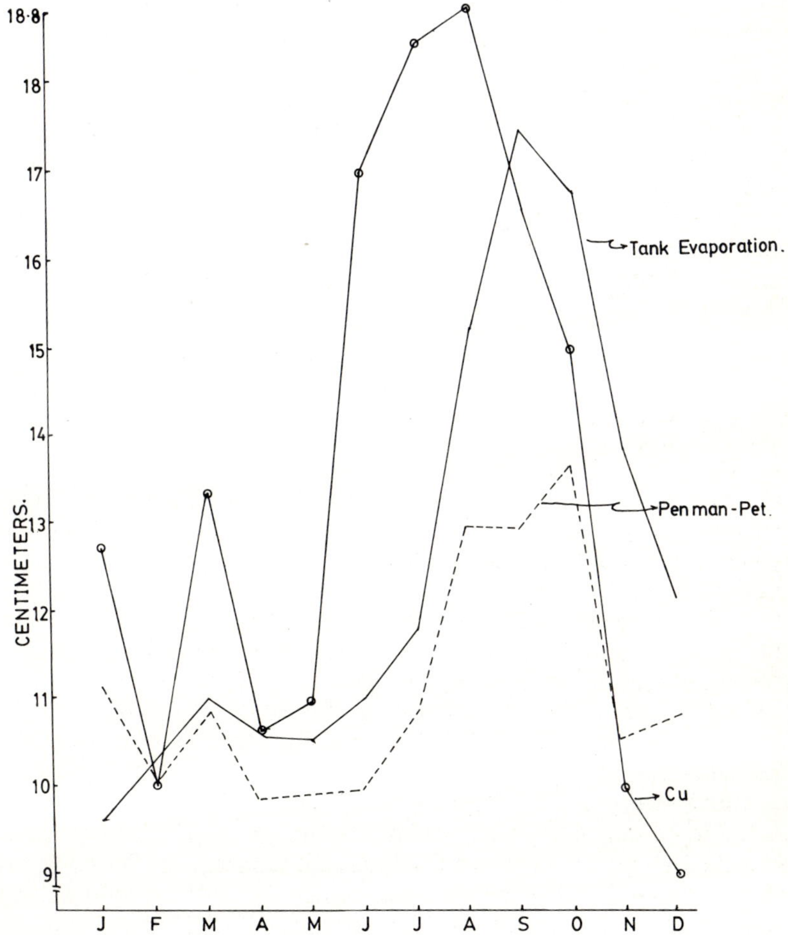

Fig. 6.2. The tank evaporation, PET (Penman) and consumptive (Cu) use (Blaney and Criddle) methods in different months in the Mosso region of Burundi, Central Africa

The empirical formulae based on climatic parameters, however imprecise, suggested a rough guide to estimate ET and schedule the irrigation of cane. Further evidence was advanced by Chang et al. (1965) who obtained a good correlation between USWB (US Weather Bureau) class A Pan and water consumption by plants. The quintessence is to irrigate sugarcane on different pan ratios using crop factors. Campbell (1967) has presented the crop factors shown in Table 6.3 for different growth stages of sugarcane.

It is important to recognize the limitations of the climatological approach to irrigation. Nevertheless, irrigation based on Class A Pan has a wide applicability (Table 6.4) and an average Pan ratio (PR) of 0.85 has been used to calculate ET (Thompson et al. 1963; Chang et al. 1968).

Turner (1990) concluded that plants mirror the water need better than the soil does. Clements (1980) had expressed a similar view and the moisture content of internodes (+5 and +6) reflects precisely the soil moisture

Table 6.3. Crop factors for different growth stages of sugarcane. (Campbell 1967)

Growth stage	ET/E Pan (crop factor)
Partial canopy	0.4–0.6
Full canopy	0.8
Peak use (boom stage)	1.0
Early senescence	0.7
Ripening	No irrigation (cut-out period)

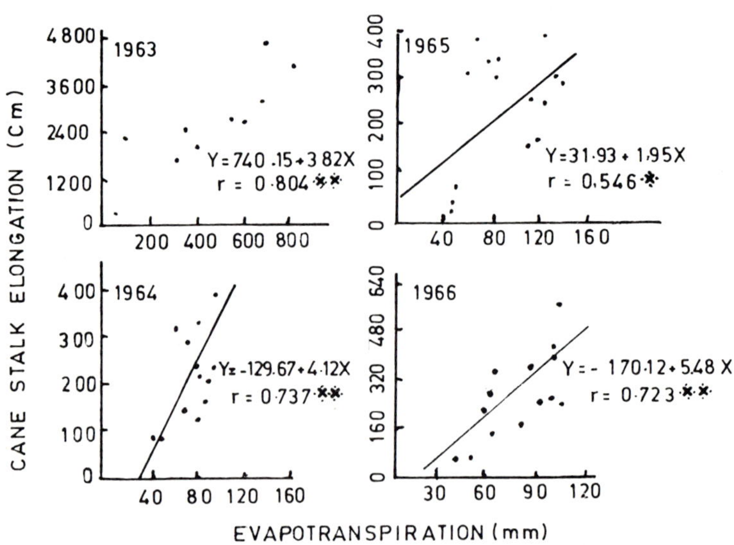

Fig. 6.3. Relation between evapotranspiration and cane stalk elongation for cane grown in a greenhouse. (Chang and Wang 1968)

Table 6.4. Effect of irrigation regimes based on PR (pan ratio) on cane yield or total water use (net irrigation and effective rainfall) in Taiwan (average for three soil types; cv. F146). (Chang et al. 1968)

1964			1965			1966		
PR	Cane yield (Mg ha^{-1})	Total water use (mm)	PR	Cane yield (Mg ha^{-1})	Total water use (mm)	PR	Cane yield (Mg ha^{-1})	Total water use (mm)
0.50	333.3	1379.5	0.4	127.8	1100.5	1.0	189.00	2000.5
0.75	384.0	1734.1	0.8	173.0	1617.5	1.25	179.56	2290.8
1.0	349.1	2090.4	1.20	146.4	2135.1	1.50	187.04	2581.0

Data for 1966 pertain to a sandy loam soil.

Table 6.5. Leaf water potential (Ψl) of potted cane with different physiological and moisture status (Inman-Bamber and Dejager 1986)

Events	Plant status	Ψl (MPa)
1	Plant extension rate reduced and youngest unfurled leaf begins to roll	− 0.8
2	Stomatal resistance starts to rise	− 0.8 to − 1.0
3	Green leaf area reduced	− 1.0 to − 1.7
4	Plant extension rate ceases and stomatal conductance reaches a minimum	− 1.3 to − 1.7
5	Youngest unfurled leaf becomes fully rolled	− 2.0
6	Least number of living leaves and eventual stalk death	− 2.8

tension and could be used as a ripening and irrigation guide. Stalk elongation and total water applied (irrigation water plus effective rainfall) showed a linear relationship as illustrated in Fig. 6.3 (Chang et al. 1968). This emphasizes the need to use a sensitive part of the plant to schedule irrigation. Recent evidence indicates that as the soil water regime declines from saturation to dryness, so stomatal conductance, net photosynthesis and leaf expansion rates are dramatically reduced (Roberts et al. 1990). The leaf rolling index is a sensitive indicator of reduced soil-water availability (Turner 1990). Inman-Bamber and Dejager (1986) reiterated that the leaf water potential (Ψ^l) in situ is the practical method for scheduling irrigation and the sequence of events is given in Table 6.5.

Irrigation should be resumed before the young leaves start to roll at a leaf water potential (Ψ^1) of -1.0 to -1.5 MPa (Table 6.5). These authors observed that N11 was better able to adapt to water stress than NCO310 in that it could adjust the leaf area more rapidly.

6.3 Water Use Efficiency

The relationship between water use and yield has been studied since the beginning of the nineteenth century and Leather (1910) proposed a transpiration ratio of 200 for sugarcane. A linear relation may be established between the DM production and the water use efficiency (WUE), which is defined as the ratio of crop yield (Y) to the amount of water depleted by the crop (ET): WUE = Y/ET. The dimensions are tons cane ha^{-1} cm^{-1} (tc $ha^{-1}cm^{-1}$) or tons sugar ha^{-1} cm^{-1} (ts ha^{-1} cm^{-1}). It is hardly necessary to emphasize that WUE has notable implications for on-farm, regional and national planning of irrigation water resources.

In their comprehensive review, Yates and Taylor (1988) gave WUE values ranging from 0.7 to 1.45 tc ha^{-1} cm^{-1} in different sugarcane-growing regions of the world (Table 6.6). It can be safely concluded that a WUE of 1 tc ha^{-1} cm^{-1} or 0.1 ts ha^{-1} cm^{-1} seems optimum. Isobe (1968) vividly

Table 6.6 WUE of sugarcane grown under field conditions in different parts of the world. (After Yates and Taylor 1988)

Country	WUE ($tc\ ha^{-1}\ cm^{-1}$)
Hawaii	0.80–1.20
South Africa	0.87–0.99
Australia	0.70–0.90
Jamaica	0.70
Argentina	0.7 –1.90
Taiwan	0.95–1.45
Florida Puerto Rico	1.30–1.80
Cuba	0.70–1.0
India (tropical/subtropical)[a]	0.73–1.08

[a] Data inserted by this author.

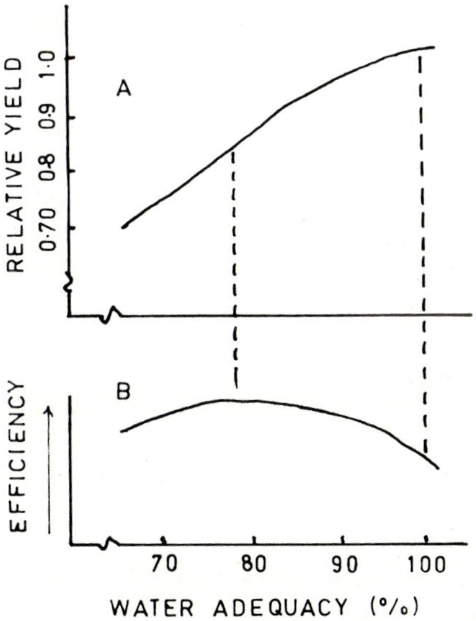

Fig. 6.4. Yield–water relationship (**a**) between relative yield and percentage water adequacy and (**b**) between water use efficiency and percentage water adequacy. (Isobe 1968)

illustrated that a maximum yield does not coincide with a higher WUE and, on the contrary, the highest WUE occurs at a point where the yield is less than the maximum (Fig. 6.4).

Several factors, e.g. control of weeds, pests, diseases and cultural practices, etc., influence WUE but the factor of paramount importance is fertilization with N and K. Viets (1962) conclusively demonstrated that fertilizers promote maximum growth but increase ET only slightly. WUE can also be improved if the water supply is adequate during the critical stages, viz. tillering and the grand growth period (2–4 months after planting/ratooning).

6.4 Irrigation Methods

General irrigation methods include surface, subsurface, sprinkler and drip systems.

6.4.1 Surface Irrigation

Water is conveyed to the point of infiltration directly on the soil surface, which is considered as the conveyance channel boundary. Kay (1990) reported that surface irrigation is the most important method of water application since more than 95% of 250 million hectares of irrigated land worldwide is irrigated in this way. Wild flooding is a very crude and imprecise method of irrigation and border strip irrigation is a more controlled process. Check-basin irrigation is suited to a uniform land slope and to soils having moderate to slow infiltration rates. In this system the field is divided into beds, usually rectangular, surrounded by bunds, and these are called check basins.

Reynolds and Gibson (1968) have detailed the irrigation system for sugarcane in Hawaii. They emphasized that the alternative systems depend on whether the need is to save water or labour, increase or protect the yield or any combination of these. An early irrigation system consisted of furrow irrigation with a fair degree of WUE and this consisted of automatic or semi-automatic water control through hydraulically actuated gates. The herringbone system consisted of a flume line laid down the slope with furrows running along the contours on both sides of the flume; this resembled a fish skeleton, hence the name 'herringbone' (Fig. 6.5a). At present furrow irrigation is commonly practised for sugarcane grown in developing countries. The length of the furrows ranges from 10 to 30 m or more depending on the soil type and slope of the land. Alternate furrow irrigation[1] can be practised which saves 25% water in addition to a 30% increase in cane yield with better sugar recovery. The serpentine method of furrow irrigation is practised for an 18-month crop (*Adsali*) in many parts of central and southern India. This method consists of forming large ridges and furrows before the boom stage (3–4 months) and the furrows are connected in the shape of a serpent (Fig. 6.5b). Water enters at one end and exits after irrigating a compact block of 0.25–0.50 ha. The advantages are that it avoids the entry of the irrigator after 3–4 months and thus prevents cane damage. Fertilization can be applied through the irrigation water (fertigation).

According to Kay (1990), the most innovative development in surface irrigation is the method of surge irrigation in furrows; this provides an opportunity to regulate the water inflow to match changing infiltration into the soil. Change in water infiltration is probably due to the breakdown of clods filling the soil pores and thus creating a smoother channel for subsequent surges.

---

[1] also called skip furrow irrigation

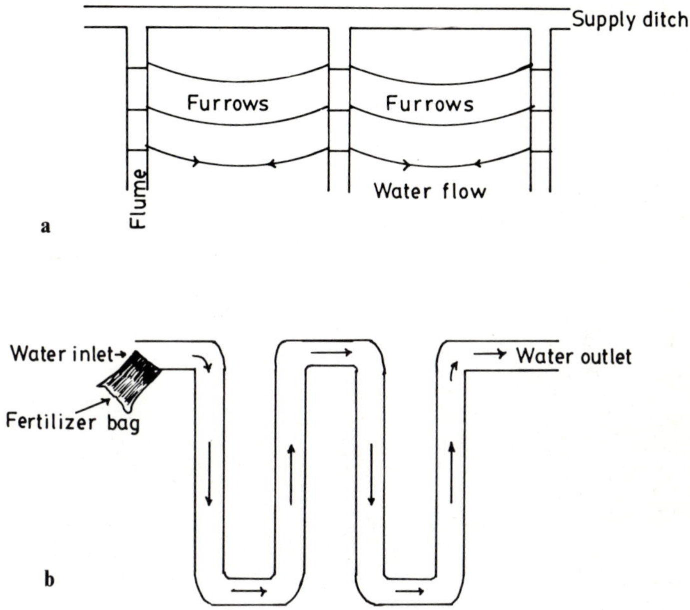

Fig. 6.5a, b. Two methods for irrigation to sugarcane **a** Herringbone system of irrigation **b** Serpentine method of furrow irrigation

6.4.2 Subirrigation

This is defined as the act of regulating the elevation of the ground-water table by artificially adding water underground (Criddle and Kalisvaart 1967). In essence it is drainage in reverse, or can be termed controlled drainage. It has wide applications in humid regions. In the Everglades in Florida, sugarcane is irrigated by subirrigation. Shih (1989) demonstrated that the seasonal ET for cane under subirrigation was 1275 mm with a yield level of 119 $Mg\,ha^{-1}$.

6.4.3 Sprinkler System

Sprinkler or overhead irrigation was in vogue in many sugarcane estates in Asia, America, Africa and Latin America by the early 1970s. This system was especially used for fertilizer application (fertigation), frost protection and temperature control. Sprinklers can apply water to the soil at rates usually equal to or less than the infiltration rate. This system is employed for soils or topographic conditions which are not suited to surface irrigation. Broadly speaking, this system is divided into (1) portable (2) semi-portable and

(3) stationary sprinklers. Boom-type sprinklers are employed for sugarcane irrigation. High-pressure nozzles (500–700 kPa) were commonly used but recent advances include the development of low-pressure nozzles (100 kPa or less) which are closely linked to the soil and cultivation practices. Still more recently, drag-line irrigation has been introduced to cane plantations (Zadrazil 1990). This is an overhead system where the sprinklers are connected by means of portable hoses and permanent or semi-permanent pipes to a pressurized water supply.

6.4.4 The Drip System

The development of drip or trickle irrigation by Symcha Blass, an Israeli engineer, ushered in a new era in irrigation practices for crops. In this system water is applied literally drop by drop to the root zone. The technique consists of laying a plastic tube of small diameter on the field near to the plant and water is delivered slowly but frequently from holes or special emitters (Shoji 1977). Shoji (1977) reported that with drip irrigation sugarcane could be planted in Hawaii on slopes of 20–30%. The chief advantages of drip irrigation include enormous water savings, improved fertilizer distribution and use of highly saline water. It is reassuring to note that while the irrigation efficiency under the conventional system is around 50% it is 70–80% for sprinkler systems and 85–95% in drip irrigation (Shoji 1977.) Nonetheless, there are some serious limitations, such as plugging of holes and emitters, high investment costs, development of dense root systems around emitters and growth of cane that is highly prone to lodging.

A simplified field layout for drip irrigation in sugarcane is shown in Fig. 6.6. At the Vasant Dada Sugar Institute, Pune, Hapse et al. (1990) observed a 14% higher yield under drip irrigation than with conventional furrow irrigation. The average WUE in drip irrigation was 1.18 t ha^{-1} cm^{-1} in contrast to 0.67 t ha^{-1} cm^{-1} in furrow irrigation. The authors concluded that there is a higher water saving in subsurface rather than surface drip irrigation.

In Swaziland, Dodsworth et al. (1990) observed 7.5% more yield in drip irrigation than in flood irrigation. The sugar yield largely mirrored the cane yield, thereby proving the superiority of the drip system (Table 6.7). The overall WUE was 16% more with drip irrigation (1.079 ts ha^{-1} cm^{-1}) than with flooding (0.928 tc ha^{-1} cm^{-1}). Paradoxically, ratoons did not respond to drip irrigation and the explanation advanced by the authors is the poor soil structure and tilth on ratooning. Miller (1990) advocated a low-head drip irrigation (2 cm or less) which has all the advantages of a standard drip system as well as being cost effective and suitable for small farms.

Batchelor et al. (1989) concluded from a series of drip irrigation trials that this system increases the cane and sugar yields and saves water. Despite these advantages, the initial costs of drip irrigation are high and Soopramanien et al. (1989) generalized that efforts are needed to increase the revenue from drip irrigation systems by intercropping sugarcane.

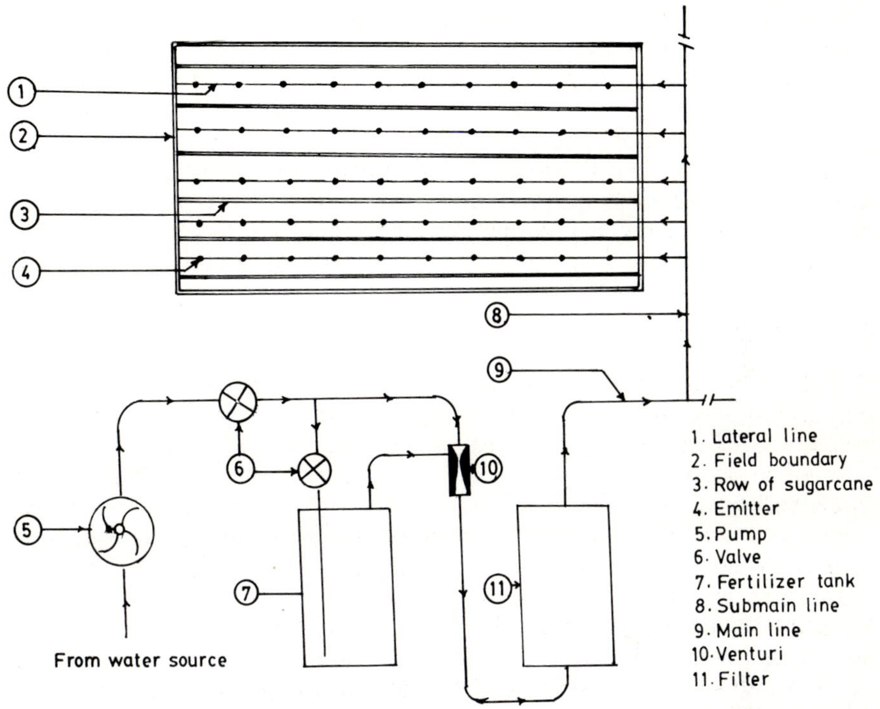

Fig. 6.6. Field layout of drip irrigation system

Table 6.7. Effect of drip and flood irrigation on cane and sugar yields and WUE. (Dodsworth et al. 1990)

Crop cycle	Cane yield (Mg ha^{-1})		Sugar yield (Mg ha^{-1})		Irrigation water applied (mm)		WUE (t ha^{-1} cm^{-1})	
	Drip	Flood	Drip	Flood	Drip	Flood	Drip	Flood
Plant crop	175.7	163.3	22.52	20.40	700	876	1.169	1.117
Ratoon I	129.3	131.7	17.29	17.11	636	769	1.041	0.924
Ratoon II	129.5	128.2	18.49	18.64	831	1209	1.009	0.772

6.5 Drought Stress

Present evidence indicates that reduced irrigation does not necessarily lower the yield or quality of cane. Varietal differences in drought resistance have already been alluded to. Drought, like other stresses, influences the endogenous levels of hormones, especially abscisic acid (ABA), ethylene, and cytokinins. Plants adjust to drought stress by the rolling, yellowing and shedding of leaves. This is accompanied by an increase in amino acids, primarily proline and betaine, and proline tests are universally accepted as an index of drought.

The management of drought stress includes selection of varieties which have drought tolerance or avoidance. Agronomic manipulations include surface mulching, 'pitting' (punching holes in loosened soil) and vertical mulching to encourage rapid infiltration of surface water. The removal of lower, partially dried leaves and mulching them on the surface ensures moisture conservation. Spraying of potash salts (KCl) at 1–2% mitigates drought and substantially improves cane quality.

6.6 Drainage

Removal of excess water on the soil surface is termed 'surface drainage', while the removal by downward flow through the soil is referred to as 'internal drainage' (Baver 1956). Drainage and irrigation are inseparable. Luthin (1957) redefined drainage as the removal of excess subsurface water in conveying devices and this is concerned with the movement of the water table in the soil. It is generally agreed that soils with an air capacity (volume of non-capillary pores) of less than 10% by volume require artificial drainage. Rough estimates show that 60 to 80 million hectares (out of 250 million hectares of irrigated land throughout the world) account for approximately 10.6% of crop land that is artificially drained (FAO 1974).

Most grasses, including sugarcane, cannot stand 'wet feet'. Irvine et al. (1984) reiterated that the yield potential of sugarcane in temperate regions is much higher than the yield presently achieved in areas where the water table is high, and subsurface drainage seems to be the remedy. Gayle et al. (1987) established a linear negative relationship between cane yield and the stress-day index due to wet conditions. The adverse effects of poorly drained soils on cane are:

1. Yellowing and curling of young leaves.
2. Reduced gaseous diffusion (influx of air and outflow of CO_2).
3. Lowered soil temperature.
4. Stagnant water and a fluctuating water table are more harmful.
5. Adventitious roots are formed.
6. Restricted root system, flattened and distorted roots and stragler roots seen (Humbert 1968). Hence, plants cannot stand drought conditions after excess water has drained away or after the water table recedes. Incidence of root rot and red rot accentuated.
7. Low sprouting and poor plant stand.
8. Internal length drastically reduced.
9. Reduced microbial activity and higher nutrient losses, especially of N, but increased content of Fe, Mn and Al.
10. Poor juice quality with a high level of invert sugars, gums and free N.

11. Profuse flowering, pithiness, a high fibre content and cane does not ripen, leading to an eventual reduction in cane and sugar yields.
12. Trafficability is reduced which is of the utmost importance in cane haulage.

A serious consequence of growing cane in waterlogged conditions coupled with high temperatures is the type of injury known as scalding. The exact cause(s) of scalding is not very well understood but a lack of O_2 in the root zone may be an important factor. A scald cane should not be immediately irrigated.

An alarming situation is the replacement of cane by rice in many countries with poor drainage. In the plains of North India, growth is restricted due to water-logging during the monsoons and to drought in the pre-monsoon period. However, drainage improves the yield of cane substantially.

Varietal differences have been noticed and varieties CP63-588, CO62175, CO775, CO7717, CO975 and CO419 are fairly tolerant to waterlogging, whilst CO1148 and CO7314 are highly susceptible. Some of these hybrids could stand flooded conditions for 2–4 weeks.

Broadly speaking, two types of drainage system, viz. surface and subsurface, are recognized. In the surface system open ditches are the cheapest but the land is lost for cultivation and is difficult to maintain due to siltation and weed infestation. For satisfactory functioning of the ditches, smoothing is carried out by forming gentle slopes towards the ditches. The drainage-by-bed-system with cambered beds is encountered in Guyana (where sugarcane lands are below sea level). These beds of 6.1 m width are separated by drains that are 0.6 m wide and 0.45 m deep (Fig 6.7). In the *Terai* region of North India (waterloggd alluvial soils) or in waterlogged Vertisols, sugarcane can be grown successfully on raised beds (30–45 cm) and the setts are planted (Fig. 6.7c) in a slanting position (45° angle). A raised bed system (Fig. 6.7d) is also used in New South Wales (Barnes 1974). A bank and furrow method has been adopted in Louisiana for low-lying, level alluvial soils with a high water table and high rainfall. This is essentially a ridge and furrow system where the furrows act as a drain.

Subsurface drainage systems include mole, tile, plastic tube and sump and pump systems. Mole drainage is perhaps the oldest system and is practised in the Everglades of Florida, in Louisiana, the West Indies, Mexico and Hawaii. Mole drains (0.75 m deep and 6 m apart) are created by pulling a torpedo-shaped object (mole) through fine-textured soils. The principal advantage is the low cost which is nearly one-tenth of the cost of tile drainage (Luthin 1957), although it is a short-lived system. Often mole drains are used for subirrigation in dry periods. Sump and pump drainage is used where water removal is not possible under gravity or where the main drains discharge into tidal rivers or oceans. This system incurs high energy costs and is almost always avoided unless essential (Humbert 1968). Tile or tube drainage can be used successfully to improve internal drainage and lasts for several years although the initial costs are high. This author can confirm that perforated clay tiles are cheap and can be used to advantage in heavy black soils.

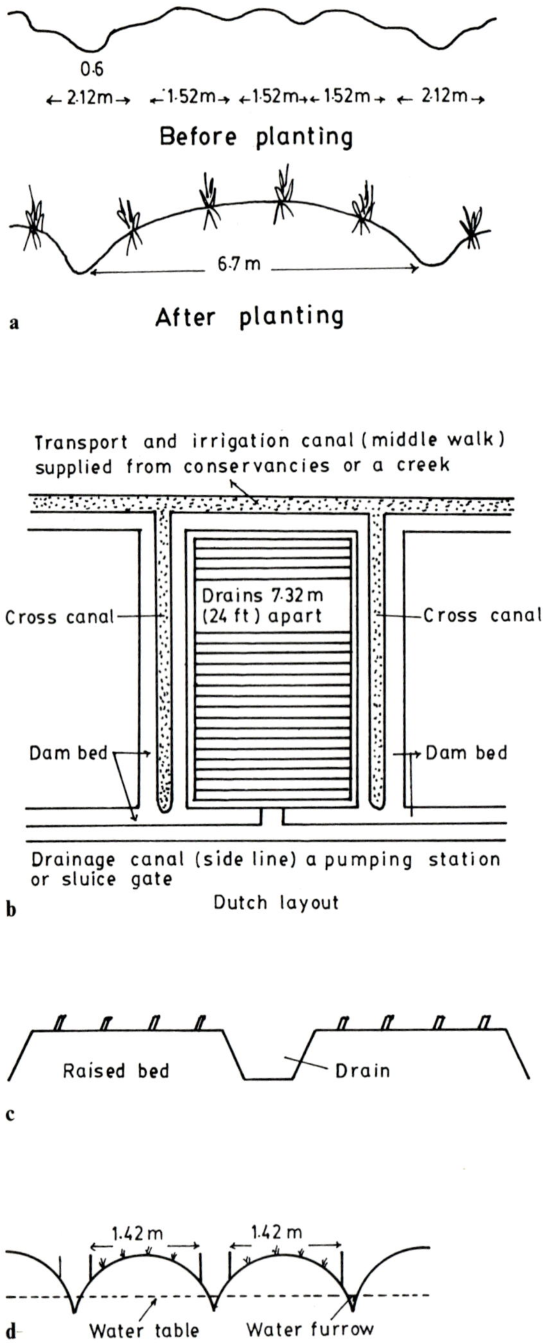

Fig. 6.7a–d. Cambered beds and raised beds. **a** A cross section of a typical cambered bed **b** The layout of a cambered bed in Guyana **c** Raised bed with slant planting **d** Raised bed system as in New South Wales. (Barnes 1974; Blackburn 1984)

One of the primary objectives of subsurface drainage is management of the water table and this is monitored with a battery of piezometers. The weight of evidence shows that sugarcane responds to subsurface drainage (Camp and Carter 1983; Carter et al. 1985, 1987; Wang et al. 1987) through improved plant stands and increased cane and sugar yields.

Wang et al. (1987) placed perforated PVC pipes at a depth of 1 to 1.4 m and obtained higher yields of 27 and 144% in first and second ratoons respectively, compared with no drainage. Carter et al. (1987) conclusively demonstrated that subsurface drainage improved the sugar yield by 339 and 791 kg ha^{-1} in two crop cycles; it also increased the number of crops harvested and enhanced the farmer's profitability. An agronomic package to improve internal drainage includes green manuring, addition of organic matter and incorporation of trash/bagasse at 40–45 t ha^{-1} (on a dry weight basis).

It has been acknowledged that sugarcane responds to good surface or subsurface drainage with improved yield and quality. Root weight is increased by more than 50% and, hence, plants can better tolerate drought conditions (Humbert 1968). The optimum water-table level is about 1 m (Hunsigi and Srivastava 1977). Thus, an integrated drainage and irrigation system ensures permanent and well-maintained sugarcane plantations.

Chapter 7 Ratooning

Ratoon cropping is defined as the cultivation of the crop growth after cane harvest, although not necessarily for grain (Francis 1989). This is essentially sequential cropping to give crop intensification in the time dimension and there is no intercrop competition. The word *ratoon* seems to originate from the Latin *ratonus*, the Spanish *retono* or the French *rejeton*. Ratoon cropping has been practised since antiquity and probably derives from early observations on the regrowth of grasses after cutting (Plucknett et al. 1970). The early record of ratooning in sugarcane dates from 1883 when an unknown Hawaiian farmer stated that 'only good ratoons pay', but the earliest ratoons seem to have been started in the Fujian province of eastern China in 1727.

It is difficult to estimate the area under ratoon, either globally or regionally. The reason is that there could be as many as six to eight successive ratoons, as in Taiwan, or just a single ratoon, as in Hawaii; the variation relates to soil problems or mechanical damage to the stools.

7.1 Advantages and Disadvantages of Ratooning

Alexander (1985) asserted that to manage only the plant crop is to ignore the finest 'botanic aspects' of the genus *Saccharum*. The essential feature of ratoon cropping lies in:

1. A shorter crop cycle.
2. Reduced costs of production, particularly of seed.
3. Better utilization of climatic conditions, especially in monsoon areas.
4. An extended milling period, because ratoons are available for grinding from the start of the sugar factories. The lower hydration levels and early flushing out of N in ratoons are possible causes for their early maturity.

In terms of energy utilization, Kishan Singh (1981) observed that production of 1 ton ratoon cane requires 89.04 million cal compared with 204.55 million cal required by the plant crop.

Ratoons have many disadvantages for they are inferior biotypes which evolved in the ecological mosaic. It was observed in South Africa (Anon 1988) that the photosynthetic efficiency of plant cane (cv. N14) was 1.8% compared with 1.7% in the first ratoon. Furthermore, at the grand growth phase the LAI of the plant crop was 5.8 while in the ratoon it was 3.0 (Irvine 1983). Build up of pests and diseases, early 'run out' of commercial varieties and poor utilization of applied nutrients, especially N, are other disadvantages. To produce a ton of cane from plant or ratoon the N required is 0.97 and 1.98 kg, respectively (Lakshmikantham 1983). In general, ratoon yields tend to be lower with a reduced juice quality compared with plant cane yields, albeit there are many exceptions to this rule. The greatest disadvantage of the ratoon lies in its restricted stubble root system.

7.2 The Root System in Ratoons

A study of the root system in ratoons is of special significance, because the response to cultivation, fertilizer application and water management depends to a great extent on the stubble roots of ratoons. The special features of the ratoon root system are:

1. Ratoons have a shallower root system than the plant crop as far as the absorption part is concerned (Peng 1984).
2. The shoots of successive ratoons originate at a higher level in the soil than those of the plant crop (Fig. 7.1) and hence ratoons have less anchorage and are more susceptible to lodging and drought.
3. Initially, ratoons start functioning on the old stubble roots, which are essentially a 'rope system' (Van Dillewijn 1952) of the plant crop. The old stubble roots are less efficient in absorbing nutrients and water because they are highly suberized.

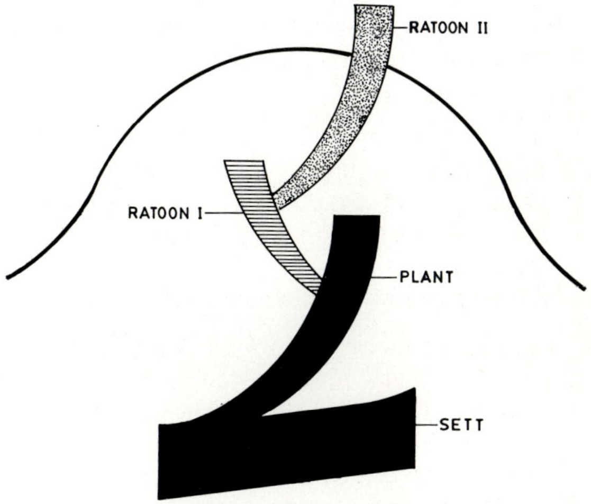

Fig. 7.1. Emergence of stalks of successive ratoons

4. At the beginning the functional, white, succulent shoot roots of ratoons are very much restricted.
5. The transition between the decay of stubble roots and the initiation of new shoot roots depends on soil and climatic conditions and can vary from 6 to 8 weeks. The old root system gradually ceases to function and decays. Under tropical conditions a ratoon will have its own root system in 2–4 months.
6. The trend of tiller production closely follows the pace of shoot root formation (Van Dillewijn 1952). Hence, the presence of late tillers is an indication of a vigorous root system.

A detailed examination of the root system of a ratoon of CO62175, grown in the red soils of Mandya (Alfisols, Xeralfs), was carried out using the method of Evans (1935), involving a jet of water. During the first 3–4 weeks after stubble shaving, more than 90% of the root system was comprised of stubble roots. The lateral spread of the stubble roots was 1.76 m while the depth was 1 m (Fig. 7.2). The formation of new white, shoot roots had just

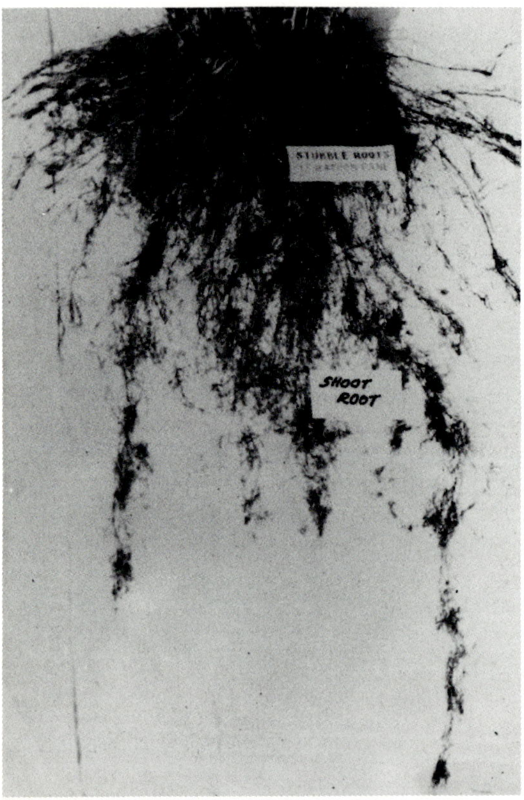

Fig. 7.2. Root system of ratoon cane (cv. CO62175)

begun. The stubble roots are greatly ramifying, darker, highly suberized and less efficient. Quoting L. D. Baver, Humbert (1968) indicated rather convincingly that absorption of ^{32}P was limited due to the presence of stubble roots. More recently, similar observations have been made at the IISR, Lucknow (Anon 1981). In the initial stages, a disadvantage of ratoons is the imbalance in the shoot root ratio but an optimal ratio of 15–20 is attained at a much later stage.

In many parts of Asia, operations like ‘off bearing’ or ‘shoulder breaking’, coupled with heavy earthing up, are advocated to facilitate the decay of stubble roots and ensure the early formation of shoot roots. The decaying stubble roots may temporarily immobilize N and reduce its availability to the crop. Immediately after the harvest of the crop, stubble shaving is carried out to encourage the development of vigorous shoot, a deeper root system and synchronous tillers (Barnes 1974). Stubble shaving is invariably practised in countries where there is manual harvesting of the crop. However, operations like ‘shoulder breaking’ and stubble shaving are not undertaken, in sugarcane-growing countries like South Africa, Australia and Taiwan. Ham (1982) even advocated that closer stubble shaving should be avoided due to the risk of a reduced yield in the next ratoon.

7.3 Fertilizer Management

7.3.1 Nitrogen

The response of ratoons to applied N is ubiquitous. Nitrogen increases leaf area, stalk elongation, tillering and yield. Research has demonstrated that ratoons are less efficient users of N than the plant crop. To produce 1 t in ratoon nearly double the dose of N is required, compared with the plant crop. The data of Chow and Samuels (1977) further confirm that NUE was 1111 kg cane kg^{-1} N applied for the first crop and the values for successive ratoons decreased to 1020 and 902 kg cane kg^{-1} N. Chapman et al. (1983) stressed that more N was required to produce a maximum sugar yield in ratoon than in plant cane. There was no residual effect of N in subsequent ratoons. Limited experiments conducted by this author in the Mosso region of Burundi (Central Africa) showed that the response to N was consistent, but was linear in the first ratoon and curvilinear in the second ratoon. The response seemed to level off at 120 kg N ha^{-1}. It was also shown by Samuels (1969b) and Gascho et al. (1986) that leaf N dropped more in ratoons than in the plant crop. Hence, Zende (1981) advocated a soil level of 40 ppm NO_3-N throughout the crop cycle to ensure an optimum yield in the plant crop and successive ratoons.

Globally, the average range of responses varied from 0.05 to 0.2 t kg^{-1} N applied. Still greater responses are anticipated in highly N-dificient soils. An increased ratoon yield following N addition is attributed to the increased number and weight of millable stalks. A reduced NUE in ratoons is ascribed to an imbalance in the shoot root ratio at the juvenile stage, delayed shoot root development and relatively inefficient stubble roots. More recently, Gascho et al. (1986) demonstrated rather convincingly that with a normal rate of N application, ratoon yields are far less than plant crop yields (Table 7.1). This is attributed to lowered internal and external NUE in ratoons compared with the plant crop. The authors calculated the external NUE in terms of kilograms of either sugarcane or sucrose per kilogram of N applied. The internal NUE was calculated as kilograms of sugarcane or sucrose per kilogram of N accumulated in the above-ground portion of the plant. However, at higher rates of N dressing, ratoons performed better than the plant crop.

Conscious efforts have been made to improve the NUE in ratoon cane by the use of biofertilizers and slow-release N fertilizers, but the results are not consistent under field conditions. Urea blended with neem cake (cake made from the seeds of *Azadirachta indica*) proved more effective (Zende 1981) and blending 300 kg N (as urea) with 140 kg neem cake (approx. 10–12% neem oil) was most effective. Point placement of urea super granules (USG) in red, sandy loam soils (Alfisol, Xeralfs) gave, over an average of four seasons, an extra 17 t ha^{-1} yield (Table 7.2).

Incorporation of intercropped legume residue, such as that from French beans, sunhemp, green gram or soyabean, improved substantially the cane yield (Table 7.3). Both the quality and quantity of biomass incorporated in the soil determine the improvement in yield of ratoon cane. The biomass incorporated in soil should be succulent, high in N content and easily decomposed. Incorporation of legume residue is associated with increased

Table 7.1. Effect of N rate on yield and its attributes in plant and ratoon canes. (Gascho et al. 1986)

N rate (kg ha^{-1})	Millable stalks ($\times 10^3$ ha^{-1})	Stalk weight (kg)	Yield (Mg ha^{-1})	
			Cane	Sugar
Plant crop				
0	34.2	0.52	24.1	3.1
56	45.3	0.56	29.7	4.0
112	41.2	0.57	27.7	3.9
224	47.7	0.59	32.7	4.4
Ratoon cane				
0	27.0	0.39	15.5	1.7
56	35.4	0.50	20.3	3.0
112	39.0	0.52	20.3	3.0
224	50.0	0.66	36.8	5.5
448	58.5	0.67	42.9	5.9
896	42.8	0.63	33.3	4.0

Note: Mg ha^{-1} = t ha^{-1}

mineralizable N (about 200 kg ha^{-1}) for up to 6 months and with increased dehydrogenase activity (30–35 μl H_2 evolved per 5 g soil).

The data in Table 7.3 reveal that there is a saving of about 50 kg N ha^{-1} and at a higher level of N dressing the effect of residue incorporation is diminished.

The response to applied N also depends on the cane cultivar; Gascho et al. (1986) observed that cv. CP65-357 gave the highest yield of cane and sucrose.

Table 7.2. Effect of urea vs USG on ratoon cane yield (Mg ha^{-1}) in cv. CO62175

Source of N (250 kg ha^{-1})	Plant crop (1979–1980)	First ratoon (1980–1981)	Second ratoon (1981–1982)	Third ratoon (1982–1983)	Fourth ratoon (1983–1984)
Urea	125	99	94	100	70
USG	124	109	112	109	102
Increase of USG over urea	1	10	18	9	32

Table 7.3. Effect of incorporation of legume residue on yield (Mg ha^{-1}) (first ratoon of cv. CO419)

Treatment	N level (kg ha^{-1})		
	200	225	250
1 Incorporation of legume residue (mean yield of four intercrops)	156	159	152
2 Entire cane	111	138	147
Increase of 2 over 1 (%)	28.8	13.2	3.3

Table 7.4. Internal and external NUE (kg ha^{-1}) of ratoon cane from four cultivars. (Gascho et al. 1986)

Cultivar	N accumulated in above-ground plant parts (kg ha^{-1})	External NUE		Internal NUE		Yield (Mg ha^{-1})	
		Cane	Sucrose	Cane	Sucrose	Cane	Sucrose
CP63-588	48	62	7	385	44	18.4	2.1
CP56- 59	56	71	10	370	50	20.9	2.8
CP65-357	63	167	24	769	110	47.1	6.9
CP68-1026	49	100	14	556	79	27.7	3.9

External N use efficiency = kg either cane or sucrose kg^{-1} N applied
Internal N use efficiency = kg either cane or sucrose kg^{-1} N accumulated in the above ground plant parts

This cultivar had the highest internal and external NUE (Table 7.4). The authors advised that the NUE of cane can be enhanced by selection under low N conditions.

7.3.2 Phosphorus

A deficiency of P results in thin, stunted stalks, reduced tillers, narrow leaves, restricted root development and slow growth. Clements (1980) stressed that ratoons require nearly double the amount of P compared with the plant crop because they have to start the root system de novo. Phosphorus is essential to hasten the formation of shoot roots and increase tillering of ratoons but its availability depends on the fixation of native and applied P. To circumvent this problem it is suggested that P should be applied in localized concentrations through carriers like compost, farmyard manure, bagasse, etc. Workers in South Africa felt that a top dressing of P over cane trash was more effective in ratoon. The role of residual P in increasing the yield of ratoon grown in Entosols (Table 7.5) has been indicated by Jaffri (1980). The explanation offered by the author for increased ratoon yield due to residual P is early shoot root formation and better mineralization of N.

Among the sources of phosphates, more concentrated triple super phosphates (44–52% P_2O_5) are used in the sugar industry. If sulphur deficiency is anticipated, then single super phosphate is used. The greater effectiveness of rock phosphates (27–41% P_2O_5) in acid soils is well recognized and under acid conditions it dissolves to give PO_3 which is protonated to become $H_2PO_4^-$. Rock phosphates with a fineness of 100 mesh (150 μm) have a higher fertilizer value in acid soils, and due to the prolonged residual effect, ratoons respond better to this P fertilizer source. The improved response is also accounted for by reduced Al^{3+} toxicity with a concomitant increase in soil pH and Ca levels.

Among the by-products of the sugar industry, filter press mud (FPM) or filter press cake serve as an important source of P fertilizer. The production of FPM is 14–18 kg t^{-1} cane with 55–65% moisture (Meade and Chen 1977). The cake is alkaline in reaction (pH 9–9.5) with a P content ranging from 0.5 to 1.2%, a Ca content of about 2% and minor elements like Mn, Fe and Zn (see Table 5.7). Research in several countries like Cuba, Mauritius and Trinidad has amply demonstrated its suitability as a source of P to sugarcane.

Prasad (1976) advocated that application of 20 t ha^{-1} FPM is optimum (Table 7.6) and the increase in cane yield ranges from 30 to 60% in plant and ratoon cane. There is an improvement in sugar yield following FPM application. Investigations in Mauritius (Ng Kee Kwong and DeVille 1988) and Cuba (Arzola and Carrandi 1982) confirm these results and application of filter cake results in an increased organic matter content of soils with enhanced N uptake by sugarcane. FPM is a better source of P than triple super phosphate and supplies both secondary and minor elements. It has also been pointed out that FPM will not result in P-induced Zn or Fe

Table 7.5. Effect of residual P on ratoon yield (Mg ha^{-1}). (Jaffri 1980)

Treatment	Ratoons			Average of 3 years
	First	Second	Third	
P residue	95.8	73.5	61.1	76.61
P residue absent	58.2	50.6	50.5	50.00

Table 7.6. Cane and sugar yields of plant and first ratoon as influenced by FPM. (Prasad 1976)

	Plant (Mg ha^{-1})		Ratoon (Mg ha^{-1})[a]	
	Cane	Sugar	cane	sugar
Control	58.7	11.98	86.0	11.99
FPM (20 t ha^{-1})	94.5	13.14	112.9	12.47
lsd 0.05	4.4	0.36	9.0	0.46
Increase of FPM over control (%)	60.98	9.68	31.27	4.0

[a] Average of three N levels

deficiencies and it is not a liming material. However, fresh filter cake should not be applied because it is a fire hazard. For better utilization of the industrial product, FPM should be well mixed with the soil prior to planting. When applied to ratoons, FPM needs to be applied in furrows and, again, should be well mixed with the soil.

In general, the response of ratoon cane to P dressing is more consistent than that of the plant crop and 1 kg P_2O_5 results in a response ranging from 0.035 to 0.075 t ha^{-1}. With regard to the efficiency of applied P, both plant and ratoon crops seem to behave similarly. The improved yield following P application is attributable to an increase in tiller production, weight per cane and final stalk population. Addition of P tended to increase the pol (polarimetrically determined sucrose) in juice and at an optimum level of P application the purity of juice was enhanced.

7.3.3 Potassium

Adequate K in the root environment is a sine qua non for successful production of ratoon. The agronomic value of K is assessed in terms of its effect on increased girth and weight per cane. A good response to applied K was observed in red, mixed red and black soils, but the response was doubtful in alluvial soils as they contain K-bearing minerals like illite. Threshold values of exchangeable K, as obtained in neutral normal ammonium acetate, for sugarcane range from 65 to 150 ppm (Duflo 1976). In South

Africa this value was raised to 225 ppm in heavy textured soils. This author has observed that exchangeable K per se is a poor index of K availability in continuous cropping, as occurs in sugarcane ratoon. Hence, a portion of the non-exchangeable but plant-available fraction, termed the 'step K' should also be included along with the labile pool to predict satisfactorily the K availability to ratoon cane (Fig 7.3). Threshold values of labile K plus 'step K' range from 400 to 600 ppm. In heavy textured soils the response to

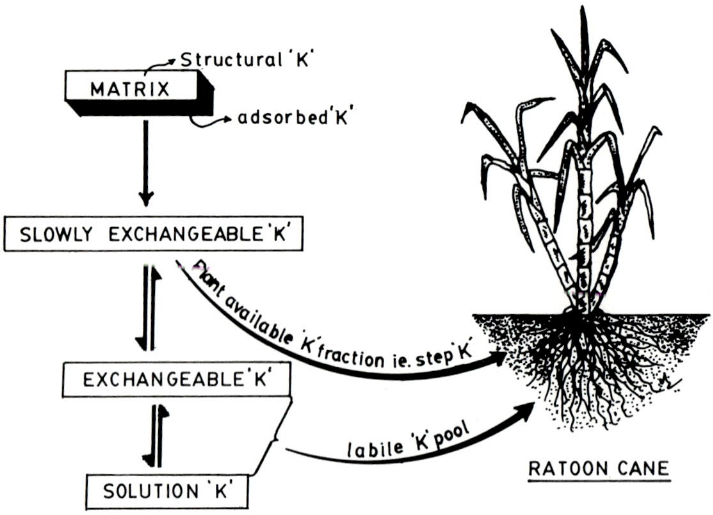

Fig. 7.3. Availability of K to ratoon cane

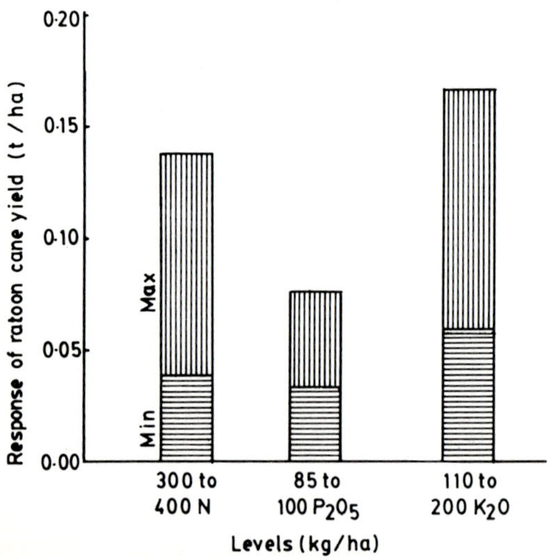

Fig. 7.4. Range of response of ratoon cane (t ha^{-1}) per kilogram of N, P_2O_5 and K_2O applied at the optimum level

applied K decreased progressively, a phenomenon which may be associated with soil compaction, decreased aeration and consequent restricted root growth. The response to applied K in red, mixed red and black soils ranged from 0.06 to 0.117 $t\,ha^{-1}\,kg^{-1}\,K_2O$ at an optimum level of 100–120 $kg\,K_2O\,ha^{-1}$.

It is reiterated that different sources of N, P and K did not lead to much difference in response in terms of cane or sugar yield. The generalized response of ratoon cane to three major elements is depicted in Fig. 7.4.

Interaction between the major elements in ratoons was noticed. Potash addition in the presence of P improved both yield and quality of cane in red, mixed red and black soils. In view of such positive interactions, Humbert (1978) suggested an optimum N:P:K ratio of 2:1:2. Duflo(1976) suggested an N:K ratio of 1:1.7. It is generalized that for ratoons an N:P:K ratio of 1:0.7:1.5 seems optimum.

7.4 Time and Method of Fertilizer Application

Studies in Australia have shown that the plant crop takes 3.5 to 5.0 months to 'close-in' compared with 3 to 3.5 months in ratoons. The fertilization schedule needs to be completed before the 'rapid close-in' period. In order to improve the fertilizer use efficiency, the time and method of application assume a greater importance. Band or point placement to the stool is recommended as this ensures easy availability of the nutrients through mass flow and diffusion. Some broad generalizations on the time of fertilizer application to ratoons are given in Table 7.7.

7.5 Secondary and Minor Elements

A large-scale response of ratoon cane to secondary elements like Ca, Mg and S has not been noticed. In some acid soils of Australia the first and second ratoons responded to lime, hydrated lime or cement and a combination of

Table 7.7. Fertilization schedules for ratoon cane

Single basal application after harvest of crop	Two equal splits: immediately after harvest and 3–4 weeks later
Medium and heavy black soils with a high CEC	Very sandy and sandy loams
Early maturing and drought-tolerant varieties	Late maturing and non-flowering varieties
High tillering varieties with initial vigour, eg. NCO310, NCO376, B37172, Q58	Low tillering varieties with low initial vigour
	Heavy rainfall areas and coastal regions

lime plus $MgSO_4$. The mean response to lime plus $MgSo_4$ was 25 Mg ha^{-1}. However, there was no response to a mixture of micronutrients (Anon. 1987a). Lime-induced Fe chlorosis is more pronounced in ratoons than in plant cane. In calcareous soils, soil application at 10 kg ha^{-1} or a foliar spray with 2% ferrous sulphate would cure Fe deficiency. In light-textured sandy soils, ratoon chlorosis is common; it is transient and due to temporary non-availability or immobility of Fe in the tissues. Thus, there are three types of Fe deficiency in sugarcane:

1. A real deficiency due to an acute shortage of Fe, but this is rare under field conditions.
2. Lime-induced Fe deficiency.
3. An imbalance in the proportion of Fe and Mn.

The last scenario occurs in the red soils of Fiji where Mn toxicity causes Fe deficiency. The optimum Fe:Mn ratio is 15:1 and the ratio reaches 1:1 when Mn becomes antagonistic, causing Fe deficiency. An excess of heavy metals, like Zn, Cu and Co, in soils would induce Fe deficiency. In the recent past, when cane has been planted or ratooned in paddy fallows, Zn deficiency has been noticed and this can be corrected by soil application of zinc sulphate at a level of 20–25 kg ha^{-1}.

The important role played by Ca and Si in sugarcane grown in Histisols (organic soils) was recognized by Clements (1980). He observed that in the absence of Ca^{2+} ratoons were weaker and the stools failed to ratoon. His experiments demonstrated that coral stone, as a source of Ca ($CaCO_3$, 92–95%), applied at 12.5 t ha^{-1} gave the maximum yield of cane and sugar. He further stressed that both Ca and Si serve as correctives for toxicities in acid soils.

Silicon plays an important part in the growth of plants but it cannot be considered as an essential element for any crop; roots absorb silicon as silicic acid (Wild 1988b). To ascertain the role of silicon, experiments were conducted by Elawad et al. (1982a) with TVA and Florida slag as sources of silicon. These materials contain 16–24% Si and 16–18% Ca as well as other essential minor elements. Irrespective of the source, silicon application to low Si soils, such as Histisols, improved the yields of cane and sugar. An increased cane yield following Si addition was attributed to increases in

Table 7.8. Response of the first ratoon to silicate rates. (Elawad et al. 1982a)

Silicate rate (t ha^{-1})	Height (cm)	Diameter (cm)	Number of stalks ($\times 10^3$ ha^{-1})	Cane yield (Mg ha^{-1})	Sugar yield (Mg ha^{-1})
0	111	2.1	66	63	4.8
5	149	2.5	85	99	7.6
10	164	2.6	94	126	9.9
15	178	2.7	106	142	11.0
20	166	2.6	77	89	6.8

plant size, plant height, stem diameter and stalk density and to improved photosynthetic efficiency (Table 7.8). The applied silicon improved leaf chlorophyll content and corrected leaf 'freckling'. The authors concluded that Si is necessary for normal growth of the plant or ratoon and that leaf 'freckling' is a symptom of its absence. The optimum rate of application of silicate material for ratoons is 5–15 $t\,ha^{-1}$ (see also Table 5.16).

7.6 Foliar Diagnosis

Crop logging can be successfully used to monitor, evaluate or schedule fertilizer application to maximize crop production. This approach has, in recent times, been contested (Elwali and Gascho 1984). Soil tests remain an excellent pre-planting practice and critical nutrient levels (CNL) are a good guide for scheduling fertilizer application to the plant crop but its reliability is questionable for successive ratoons (Elwali and Gascho 1984). Recent research seems to favour nutrient ratios rather than the CNL approach. Hence, the diagnosis recommendation integrated system (DRIS) seems to have the edge over the Clement's/Hawaiian crop log approach. It has been pointed out by Elwali and Gascho (1984) that DRIS is more suitable for successive ratoons as a foliar diagnostic technique and is a better guide to fertilizing cane. They obtained higher cane and sugar yields when the fertilizer dose was based on DRIS rather than on soil tests or the CNL approach (Table 7.9). This has been attributed to the late application of both major and minor elements, as revealed by DRIS values.

7.7 Quality: Ratoon vs Plant Cane

Ratoons mature earlier than the plant crop due to early dehydration of the tissues and to 'flushing out' of N. This is the reason why ratoons are preferred for milling early in the season. The quality estimates of plant and

Table 7.9. Effect of soil testing, the CNL approach and the DRIS system on ratoon cane and sugar yields. (Elwali and Gascho 1984)

Method/approach	Yield ($Mg\,ha^{-1}$)	
	Cane	Sugar
Soil test	75.4	7.9
Foliar diagnosis: CNL approach	74.3	7.7
Foliar analysis: DRIS	85.9	9.4
lsd 0.05	5.3	0.7

Table 7.10. Pol ratio (TC/TS) in plant vs ratoon cane

	Variety			Planting/ratooning			Age at harvest (months)		
	CO62175	CO419	B37172	Aug.	Nov.	Feb	10	12	14
Plant crop	9.31	7.31	7.10	9.51	7.07	5.86	7.41	7.16	7.70
Ratoon I	6.07	4.50	3.76	4.90	4.33	4.96	4.51	4.62	5.03

ratoon cane (Table 7.10) are made by the pol ratio, that is, tons cane/tons sugar (TC/TS). Obviously, the lower the pol ratio the better the quality and values can range from less than 5 to over 15. Table 7.10 indicates that ratoons of three varieties at different ratooning dates and harvest ages matured earlier than the related plant crop.

7.8 Population Dynamics

Sugarcane ratoon inherits about 7% as stubble from the preceding crop and about 3% as root mass (Shrivastava et al. 1985). Barnes (1974) pointed out that only a good plant crop can produce good ratoons. In many tropical areas of Asia the population dynamics are such that starting with more than 30 stalks m^{-2} there will be an exponential drop, stabilizing at 10–12 stalks m^{-2}. This decrease has been noted by several workers (Tang and Ho 1968) who have observed 50 to 60% mortality in the formative phases. Consequently, early earthing up (at 2–2.5 months) is advocated to arrest formation of late tillers.

The yield parameters of ratoons are the stalk density, stalk diameter and weight per cane, in that order, and together these account for about 80% of yield variation. Barnes (1974) emphasized that stubble shaving and earthing up provide stout and vigorous sprouts and are effective against frost injury.

7.9 Cultural Requirements

A great diversity is observed with regard to the cultural requirements of ratoons. In India, Kenya and Uganda stubble shaving and 'shoulder breaking' are advocated while these operations are totally dispensed with in countries like South Africa, Australia, Taiwan and many Latin American countries. It is reasonable to believe that in regions where manual harvesting is carried out, stubble shaving has a distinct advantage in promoting uniform, stout and vigorous shoots. In Swaziland and South Africa, ripping and chiselling did not result in a yield advantage. In fact, ripping resulted in a significant yield reduction (Leibbrandt 1984). In Cuba and Brazil deep cultivation, to a depth of 30 cm, was beneficial for it reduced bulk density and improved the yield. It is a common observation that where heavy machinery is used successive ratoons leave the soil compacted with an increased bulk density and reduced porosity. Humbert (1968) indicated that root development was gradually retarded as the bulk density increased from 1.1 to 1.6 $g\,cm^{-3}$. Root proliferation and extension practically ceased at a bulk density of 1.9 $g\,cm^{-3}$. In such compacted soils the O_2 potential was lower and there was consequent reduction in uptake of water and nutrients. Juang and Uehara (1971) convincingly demonstrated reduced uptake of P (^{32}P) and K (^{86}Rb) as the bulk density increased in a clay loam soil (see Table 3.3). It is suggested that in heavy soils 'shoulder breaking' or 'off barring' helps to 'open the soil', permitting better aeration and water infiltration, facilitating the decay of old stubble roots and promoting early shoot root formation. The bottom line is that under most soil conditions, subsoiling in the inter-rows of ratoons is not worthwhile and a conventional plough depth of 20–25 cm is quite adequate (Moberly 1969). Conservation tillage has been recommended in Australia, while minimum tillage through chemical ploughing using glyphosate was advocated by Iggo and Moberly in South Africa (1976). With regard to the harvesting time of subsequent ratoons, there is a general consensus that they should be harvested early. Ambient temperature, humidity and other parameters have a profound influence on ratoon yield and quality.

With respect to spacing, a relatively closer spacing of 0.5 m is better than 1.0 or 1.5 m to obtain a higher biomass and cane and sugar yields. On the other hand, closer spacing creates problems with weed control (Shih and Gascho 1980) and working with machinery.

7.10 Gap Filling

This is the most important operation for successful ratooning in sugarcane. Gaps in ratoons arise due to poor sprouting, mechanical damage to stubbles roots and from pests and diseases. Experience gained in several countries has shown that nearly 25–30% of the area needs to be gap filled. Poor cane ratooning in Taiwan is attributed to the failure of buds to germinate due to the presence of nymphs of *Mogannia hebes*, which may produce a toxin called bud germination inhibitor (Hsia and Ou-Yang 1968). For successful gap filling conditions of supersaturated soil moisture and a high ambient humidity are desirable. Research by Prasad et al. (1981) reveals that there is a significant reduction in yield when gaps exceed 20% (Table 7.11). It is worth replanting if the gaps exceed 50%. In another study, Kanwar and Sharma (1981) observed that gap filling the ratoons with pre-germinated setts, raised from single buds in polythene bags, ensured quick establishment with higher tiller and stalk populations. Horizontal planting was better than vertical planting.

Table 7.11. Effect of gaps on the performance of a ratoon stand (cv. CO1148). (Prasad et al. 1981)

Intensity of gaps (%)	Initial clumps ha^{-1}	Millable canes per clump	Millable canes ha^{-1} ($\times 10^3$)	Yield ($Mg\,ha^{-1}$)	
				Cane	CCS[a]
0	33.3	3.24	107.9	78.6	8.34
10	30.0	3.51	105.1	77.4	8.32
20	26.7	3.78	100.7	70.8	7.56
30	23.3	3.84	89.5	66.1	7.06
40	20.3	3.88	77.7	60.5	6.46
50	16.7	4.02	67.1	52.7	5.79
lsd 0.05	—	0.26	6.4	6.1	0.85

[a] CCS, Commercial cane sugar.

Under adverse climatic conditions, like low temperatures and frost, stubble sprouting is enhanced by spreading transparent polythene over the soil. This resulted in an increased number of shoots and stalks, with a consequent increase in yield. Similar observations were made by Millard (1974) under South African conditions, where plastic mulching (20 μm thick) resulted in increased cane and sugar yields of 25 and 2.8 $Mg\,ha^{-1}$, respectively. A polythene mulch effectively controlled weeds, while a liquid mulch is suggested to be cost effective and probably helps planted and successive ratoons to improve yield and quality (Rau and Millard 1975).

7.11 Number of Ratoons

In the Fujian province of eastern China there are ratoons which are nearly three centuries old. Researchers in South Africa advocate the use of multiple ratoons to 'reap rich rewards'. In Venezuela 9–20 ratoons and in Taiwan 7–8 ratoons were common. In Australia 2–3 ratoons are grown, while Hawaii adopts single ratooning. In Pakistan a single ratoon is used since the yield is reduced by 40–60% compared with the plant crop. But even a single ratoon is improbable in saline sodic or acidic soils. Interestingly, Blackburn (1984) has observed that the number of ratoons are controlled by a statute or by legislation in Barbados, Queensland, Java and Taiwan. Multiple ratooning can probably be practised in well-drained, deep soils with a high native fertility. This calls for careful gap filling, judicious and timely application of organics (including green manuring), trash mulching to conserve soil moisture, reduction of soil compaction by avoiding cane loading in the field, use of oxen and cultivation of suitable varieties. This is corroborated from a study made by this author in the Mosso region of Burundi (Central Africa); the results indicated that relatively more ratoons can be taken in heavy alluvial soils than in light-textured Ferrisols (Table 7.12).

Table 7.12. Performance of plant crop and ratoons in two soil types in Mosso, Burundi (cv. NCO310)

Soil type	Crop	Age at harvest (months)	Yield ($Mg\,ha^{-1}$)	Number of stalks m^{-2} at harvest
Ferrisols (light textured)	Plant crop	14	155	13.9
	First ratoon	14	102	14.5
	Second ratoon	14	90	14.4
	Mean	14	116	14.3
Alluvions (deep and heavy textured)	Plant crop	14	134	15.6
	First ratoon	14	136	12.7
	Second ratoon	14	178	17.7
	Mean	14	149	15.3

However, Peng (1984) has asserted that in recent years there has been a tendency to reduce the number of ratoons to no more than 2 due to an evident build-up of injurious pests, diseases and weeds. To maintain healthy plantations it is recommended that no more than 2–3 ratoons are raised in any particular sugarcane-growing region of the world.

7.12 Ratooning Power of Cane Cultivars

Good ratooning power of cane cultivars is an essential prerequisite for success but the question arises as to what makes one cultivar a better ratooner than others. The significance of ratooning power has been recognized since the 1930s and while varieties like CO312, CO205 and CO285 were released for general cultivation, highly adoptable varieties like CO419 and CO421 found a place of prominence in Asia and Africa. The experience of this author has shown that B37172, NCO310 and NCO376 are good ratooners but that Q49 is not a good ratooner. It is recognized that varieties which possess a high stalk density tend to be better ratooners (Anon. 1987b). The decline in yield in older ratoons was due to a decrease in the stalk population while the stalk weight remained fairly constant. To quantify ratoon productivity on Jamaican sugar estates, the ratoon performance index (RPI) was developed (Shew 1989). This denotes the accumulated yield decline between a reference yield (yield of plant crop or average yield of plant crop and first ratoon) and the yield of four successive ratoons. Realizing the yield decline as a common phenomenon in successive ratoons, Brzesowsky (1986) proposed yield declining factors (FY). These are:

FYP = 0, FYR1 = 0.9, FYR2 = 0.85, FYR3 = 0.75, FYR4 = 0.65, FYR5 = 0.55 and FYR6 = 0.50,

where P represents the plant crop and R1, R2 etc. are the first and second ratoons and so on.

Table 7.13. Ratooning power and cane yield ($Mg\,ha^{-1}$) of cane cultivars, determined from long-range ratooning trials at Mackay, Australia. (Anon. 1987b)

	Plant crop	Ratoon I	Ratoon II	Ratoon III	Ratoon IV	Ratoon V	Mean
[a] Q58: Yield	81.3	82.9	73.6	66.9	59.5	57.3	70.9
RPI	—	—	10.0	5.5	3.7	3.3	7.6
NCO310: Yield	73.3	70.9	67.7	62.5	52.8	56.3	64.5
RPI	—	33.3	14.2	7.14	3.7	4.3	8.3
Q68: Yield	81.3	71.9	60.5	47.2	37.3	34.3	56.1
RPI	—	9.1	4.0	2.4	1.8	1.7	3.3

[a] The reference yield for Q58 is the mean yield of the plant crop and first ratoon (82.1 $Mg\,ha^{-1}$). Original figures in t/acre converted to $Mg\,ha^{-1}$ ($\times 2.47$).

An attempt has been made to evaluate the ratooning power through the ratoon power index (RPI), which is defined as the reciprocal of the ratio of yield difference between the plant crop (average of plant crop and first ratoon) and successive ratoons. It is discernable from Table 7.13 that Q58 and NCO310 are better ratooners than Q68.

Ricaud and Arceneaux (1986) showed that the rates of decline in yield with ratooning differed among varieties. Cultivar CP48-103 could be ratooned up to fourth ratoon, while L60-25 and L62-96 could be used for only two ratoons. Biochemical studies of stubble indicated that a high proportion of reducing and non-reducing sugars and mineral elements, including minor elements, ensure a better ratooning power of cane cultivars (Shrivastava et al. 1981).

7.13 Trash Management

Trash handling is an important operation in ratooning. Trash is invariably burnt after the harvest of the previous crop (Fig. 7.5). Thompson (1966) has thoroughly investigated the effect of trashing (putting trash on the soil surface or trash mulching) on the yield and quality of cane. Proponents of trashing suggest that it leads to higher cane and sugar yields and to better control of weeds and early shoot borers. In Taiwan, Tang and Ho (1968) obtained 3 and 4% increases in the cane and sugar yields following trashing but the disadvantage of it is that it allows the build-up of pests and diseases. Raking and alignment is labour intensive and an impediment to surface irrigation. Trash does not easily decompose due to the high content of lignin and hemicelluloses. Moberly and McIntyre (pers, comm.) observed that

Fig. 7.5. Burning of trash after the harvest of previous crop

Table 7.14. Effect of water conservation treatments in areas with low rainfall on cane and sugar yields of two ratoons. (Moberly and McIntyre pers. comm.)

Treatment	Cane yield ($Mg\,ha^{-1}$)		Sugar yield ($Mg\,ha^{-1}$)	
	First ratoon	Second ratoon	First ratoon	Second ratoon
Trash blanket	62	54	8.2	6.8
Basin Lister	64	50	8.7	6.5
Bare soil	62	42	8.0	5.4
Mean	63	49	8.3	6.3
lsd 0.05	8	10	0.95	1.4

nearly 40 years of research have gone into trash management but its contribution towards nutrient build-up or improved soil properties is negligible. However, one distinct advantage of aligning trash in the rows is the conservation of soil moisture (Table 7.14), which is reflected in increases in stalk density, cane yield and sugar output.

7.14 Water Requirement

Campbell (1967) observed that to produce 1 kg cane, or 0.1 kg sugar, approximately 100 kg water is required. The consumptive use of sugarcane ranged from 1267 mm in South Africa, through 1499 mm in Mauritius to 1522 mm in Australia (Thompson 1966) (see also Table 6.1). We have observed the modulation of ET values for sugarcane grown under conditions of a high water table. A direct relationship was observed between total DM production and water consumption, i.e. 7–9 $g\,l^{-1}$ (Bull and Glasziou 1975). As stated earlier, ratoons are more prone to moisture stress than is the plant crop. It is therefore recommended that the plant crop be irrigated at 50% available soil moisture (ASM), while ratoons should be irrigated at 75% ASM in sandy loams and medium black soils. Workers in Australia found that ratoons can be irrigated at a Pan ratio of 0.85. This author observed that when ratoons were irrigated at 74 mm cumulative Pan evaporation (CPE, 14–15 days interval) in conjunction with use of trash mulch at 3 $t\,ha^{-1}$, yields were on a par with those obtained when frequent irrigations at 19 mm CPE (4 days interval) were used. Similarly, double-row planting with trash mulch (3 $t\,ha^{-1}$) in the skipped area is superior to other methods of irrigation as evidenced by the higher cane yield and enormous savings in water (Table 7.15).

Once again the efficiency of trash mulch in moisture conservation is evident. More recently, Shih (1989) observed a linear relationship between yield and ET. Ratoons responded more to drip irrigation than to subsurface

Table 7.15. Effect of irrigation method on first ratoon (cv. B37172) (Hunsigi unpubl.)

Treatment	Spacing (cm)	Yield ($Mg\,ha^{-1}$)	Pol in juice (%)	Number of plants m^{-2}	Weight per cane (kg)	Total water consumed (mm)
1 Alternate furrow irrigation	90	110	18.66	15.72	1.00	1202
2 Skip furrow irrigation	90	101	17.03	14.14	1.01	1193
3 Double-row planting with 120-cm skipped area plus trash mulch at 3 $t\,ha^{-1}$	60	129	18.12	16.72	1.05	1193
4 Normal practice	90	111	19.08	16.50	1.25	1541

Effective rainfall was 435 mm with water table at 1 m depth

Table 7.16. Effect of drip and subsurface irrigation on plant and ratoon canes. (Shih 1989)

Crop	Irrigation method	Cane yield ($Mg\,ha^{-1}$)	ET (mm)	Standard Pan evaporation (mm)
Plant crop	Drip	147	1300	
(335 days)	Subirrigation	137	1399	1476
First ratoon	Drip	151	1437	
(316 days)	Subirrigation	132	1344	1544
Second ratoon	Drip	130	1231	
(333 days)	Subirrigation	87	1081	1477

irrigation (Table 7.16). In tropical India, the WUE (of cv. CO7219) was 1.18 $t\,ha^{-1}\,cm^{-1}$ for drip irrigation compared with 0.67 $t\,ha^{-1}\,cm^{-1}$ for furrow irrigation.

7.15 Management of Weeds, Pests and Diseases Associated with Ratoons

From more than 200 species of weeds infesting sugarcane fields (Peng 1984) some dozen species, like *Cynodon dactylon, Panicum repens, Imperata cylindrica, Ipomea hardwichii, Cyperus rotundus* and *Sorghum halepense*, are known to be noxious and to persist in subsequent ratoons. Ratoons require an early weed-free environment (for 60–75 days) before the 'close-in' period. Pre-emergence sprays of atrazine/trifluralin/diuron effectively control the weeds. Diseases associated with the ratoons are smut (*Ustilago scitaminae*), grassy shoot disease (GSD), ratoon stunting disease (RSD), red leaf spot (*Helminthosporium saccharii*) and ratoon chlorosis. Diseases like smut, GSD and RSD can be controlled by heat therapy. Pests specific to ratoons are shoot, root and top borers, *Pyrilla*, black bugs, white fly, scale insects and mealy bugs. Scale insects can be controlled by application of malathion at 1 kg active ingredient (a.i.) ha^{-1} in 800–1000 l of water. To control *Pyrilla*, 5% BHC dust at 20–40 $kg\,ha^{-1}$ can be used and to control top borers carbofuron should be applied to the soil at 1.0 $kg\,a.i.\,ha^{-1}$. The importance of integrated pest management, including cultural, mechanical and biological practices, is stressed.

7.16 Economics of Ratooning

Information on this aspect is sketchy. Blackburn (1984) indicated that the average cost of cultivation per ton of plant cane in Trinidad and Jamaica comes to around £1.96 compared with £0.86 for ratoon cane. A sample survey by Sastry and Ramanna (1978) in some irrigated tracts of peninsular India showed that a 12-month plant crop required 482 man-days compared with 295 in ratoons. Additionally, there is a clear saving in seed material of 5–7 t ha^{-1}. The operational costs of plant and ratoon crop was 67.3 and 54.4%, respectively. In essence, there is a saving of about 13–15% in the total cost of production due to ratooning.

7.17 Effect of Growth Regulants on Ratoon Yield and Quality

Vlitos (1974) has reviewed the role of growth regulants in sugarcane and observed that four types, namely auxins, gibberellins, cytokinins and growth inhibitors, occur naturally in cane. Peng (1984) presented the recent trends in the use of these regulants to improve sprouting. The earliest observation was the breaking of apical dominance by the endogenous auxin, indole-3-acetic acid (IAA). Early attempts by Kanwar and Kaur (1977) showed that spraying 0.25% emisan or 0.5% agallol (a mercurial compound) on to the stubble, there was an improvement in sprouting, stalk density and yield. More recently, the growth regulants tested on ratoons include gibberellic acid (GA), IAA, indole-butyric acid (IBA), tri-iodo-benzoic acid (TIBA), ethrel (2, chloroethyl phosphonic acid), and chlorocholine chloride (2, chloroethyl-trimethylammonium dichloride), popularly known as Cycocel (CCC). Results have demonstrated that these chemicals improve sprouting, the millable cane population at harvest and ratoon yield (Table 7.17).

The physiological explanation is that the higher concentration of applied exogenous auxins induces the synthesis of ethylene in treated stubble, which destroys endogenous inhibitors (Peng 1984).

These regulators diminished apical dominance with a consequent increase in the number of ratoon sprouts. Studies conducted by this author further confirm these trends (Table 7.18 and Figs. 7.6 and 7.7).

From the presently available literature it is deduced that the promising growth regulants are CCC (10 kg ha^{-1}), GA (100 ppm) and ethrel (500 ppm) which are to be sprayed on ratoon stubble. These regulants improve the yield and quality of ratoon cane and can be mixed with herbicides like atrazine/diuron and sprayed on to the stubble with no adverse effects. The application of these growth regulants seems to be a remedy for adverse

Table 7.17. Effect of some growth regulants on ratoon yield. (Peng 1984)

Chemical	Yield (Mg ha^{-1})
Ethrel at 500 ppm	49.58
IBA at 110 ppm	46.50
TIBA at 50 ppm	43.75
Control	8.33

Fig. 7.6. Effect of growth regulators on first ratoon of sugarcane (ethrel)

Fig. 7.7. Effect of growth regulators on first ratoon of sugarcane (GA)

Table 7.18. Effect of growth regulants and trash mulch on millable canes and yield (cv. CO419). (Bhale and Hunsigi unpubl.)

Treatment	Cane yield ($Mg\,ha^{-1}$)	Yield increase over control ($Mg\,ha^{-1}$)	Millable canes (10^3) ha^{-1}
Trash mulch ($3\,t\,ha^{-1}$)	90	1.0	115
Gibberellic acid (100 pm)	112	23.0	128
Ethrel (500 ppm)	104	15.0	118
Emisan (0.2%)	91	2.0	117
Control (water spray)	89	—	112

Note: First ratoon grown in red sandy loams (Alfisols) and chemicals sprayed once on to the sprouts

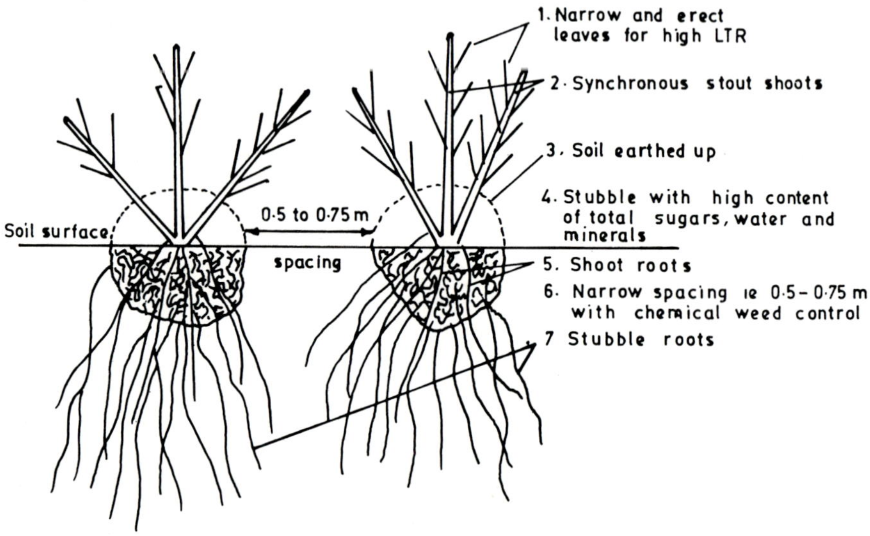

Fig. 7.8. A model of ratoon cane

climatic conditions like frost and cold. In addition, they seem to enhance the rate of respiration of stubble with an increase in sprout and stalk populations.

7.18 Modelling Ratoon Cane

Future strategies should include early and effective sprouting, quick shoot-root formation, greater ratooning power, a higher crop growth rate (more than $75\,g\,m^{-2}\,day^{-1}$), development of narrow and erect leaves with an LAI of no less than 6 at 'rapid close-in' and with a high light transmission ratio (LTR) of more than 30% to ensure reduced tiller mortality and a higher stalk density at harvest (Fig. 7.8). Ratoons tend to have a lower partition coefficient than plant cane. Row spacing of 0.5 to 0.75 m coupled with increased water and fertilizer application and a reduced crop extinction coefficient (K) assure a higher photosynthetic efficiency of the crop canopy and a consequent increase in yield and quality of ratoon cane (Fig. 7.8).

Chapter 8 Weeds, Pests and Diseases

8.1 Weeds

Weeds, innocuous and yet undesirable plants, need to be managed effectively in the widely spaced sugarcane crop. According to Peng (1984), 200 weed species are known to infest sugarcane fields and 30 are of economic importance. The most common families are Cyperaceae (35%), Umbelliferae (30%) and Gramineae (18%). Quoting FAO figures, May (1980) has listed that globally there are 10 noxious weed species and 7 which are difficult to control. These are *Cyperus rotundus*, *Cynodon dactylon*, *Echinochloa crusgalli* (sensitive to dalapon), *Sorghum halepense*, *Imperata cylindrica*, *Panicum maximum* and *Eleusine indica*. Other weeds of economic importance in sugarcane fields include *Ageratum conyzoides*, *Panicum repens*, *Tridax procumbens*, *Portulaca oleracea* and *Digitaria sp.* In Asia and the Pacific region, *Ipomoea hederacea* has reduced the cane yield by up to 25% by twining around clumps. This results in bending of the cane which reduces the height and girth and interferes with harvest operations. A similar effect of *I. hardwichii* has been seen in India and Taiwan. The root parasite *Aeginetia indica* takes food from the sugarcane through its haustorium. This parasite produces enzymes which reduce sucrose to glucose. Peng (1984) observed that healthy plants contain 13.08% sucrose compared with 5.09% in affected plants and he observed that NCO310 is highly resistant to this parasite. Thus, the losses caused by weeds are enormous as they compete for light, moisture and soil nutrients in addition to serving as alternative hosts for many pests, diseases, viruses and mycoplasmas.

Sugarcane requires a weed-free environment for the first 2 months when tillering occurs. Peng (1984) asserts that the first 3–6 weeks are critical for autumn-planted cane. It can be generalized that the first 90–100 days are critical before the 'rapid close-in' of the cane canopy. Furthermore, frequent irrigations accentuate weed infestation, and weed control in cane is achieved by mulching with bagasse and trash at 3–5 $t\,ha^{-1}$ in the cane rows. Incidentally, mulching raises the soil temperature and speeds up germination. An opaque polyethylene mulch has been successfully used in spring-planted cane in Taiwan and the yield increase of 15–30% is attributed to the rise in soil temperature and to effective weed control (Peng 1984). Flooding is employed to control weeds in sugarcane–rice intercropping. Effective weed control can also be achieved by intercropping with 'smothering crops' (spreading habit) such as French beans and green gram. Lucerne was not effective in controlling weeds (Table 8.1).

Table 8.1. Effect of intercrops on weed count, weed weight and N uptake by weeds in sugarcane at the 5-month stage and on cane yield at harvest (cv. CO62175, plant crop). (Prakash and Hunsigi unpubl.)

Treatment	Weed count m^{-2}	Dry weed weight ($g\ m^{-2}$)	N uptake by weeds ($kg\ ha^{-1}$)	Cane yield ($Mg\ ha^{-1}$)
Cane + Sunhemp	118.6	67.44	5.37	155
Cane + French beans	133.0	78.59	6.28	144
Cane + green gram	129.0	82.89	6.63	151
Cane + lucerne	127.9	106.70	8.54	142
Cane entire	150.0	103.67	8.29	128

Table 8.2. The main herbicide groups. (Peng 1984)

Contact	Translocative	Residual
Applied to the foliage; do not move from the site of application; e.g. substituted phenols (PCP), oils, arsenic compounds, diquat and paraquat	Applied to the foliage and move from the site of application; systemic action; move with metabolites through phloem; e.g. chloro-phenoxy compounds, MCPA, 2,4-D, 2,4,5-T dalapon, TCA amitrole, 2,3,6-TBA	Soil 'herbicides' or soil sterilants applied to the soil, absorbed through roots and move through xylem in transpiration stream; applied pre-planting/pre- or post-emergence. Translocative or residual since persists in soil; e.g. chloro-phenoxy compounds, MCPA 2,4-D, 2,4,5-T, symmetrical triazines, (atrazine, simazine, ametryne), other triazines (metribuzin, cyanazin), substituted ureas, (diuron, monuron, linuron norea, trifluralin, amiben, asulam)

King (1966) listed several crops, viz. potato, sweet potato, tomatoes, vegetables, peanuts, soyabeans, grain sorghum and sugarbeet, that can be used as intercrops and noted that taller crops compete better with weeds but the reduction in cane yield is 11–16%.

Both in the developed and developing world herbicides are increasingly used on sugarcane, either singly or in combination with mechanical or cultural operations. Peng (1984) states that about a dozen compounds are used in sugarcane cultivation and these can be classified into three broad groups in accordance with their mode of action as shown in Table 8.2.

Interestingly, Peng (1984) observed that toxicity to the cane is less at early stages when transient sett roots are formed. However, toxicity is much higher at the 3–5 leaf stage when shoot roots have developed and can absorb the soil herbicide. It is therefore postulated that pre-emergence application of herbicide is safer than post-emergence application due to the formation of shoot roots at later stages. Similarly, ratoons are more tolerant to soil herbicides than is the plant crop since shoot roots are formed much later after the decay of old stubble roots. Contrarily, pre-raised plantlets (*rayungans*) are more susceptible to herbicides as the shoot roots have already developed. Varietal differences have also been observed with regard to herbicidal toxicity and NCO310 is most tolerant to diuron, while F150 is most susceptible.

The residual toxicity of these herbicides applied pre- or post-emergence at 2–3 kg ha^{-1} is rarely noticed since sugarcane is of long duration (12–14 months or more) and herbicide residues are broken in a couple of weeks. Further, soil is worked during intertillage operations which help in the degradation of herbicide residues and, hence, toxicity may not be of much consequence.

The most commonly used herbicides in cane culture are: asulam, atrazine, TCA, metribuzin, diuron, cyanazin, ametryne, trifluralin, alachlor, hexazinone, paraquat and phenoxy-acetic compounds. The efficacy of these herbicides depends on the soil and climatic conditions, the cane variety and type of herbicide used. Rochecouste (1959) suggested that the evaluation of herbicides should be based on a visual scoring from 0 (no effect) to 8 (complete check and 100% weed kill). Similarly, the assessment of cane injury from herbicides can also be made by a visual rating system (0 to 10 points scale).

It is generalized that herbicides like diuron, atrazine or 2,4-D applied at 2–2.5 kg a.i. as a pre-emergence spray are effective and economical in cane fields. Application of diuron plus 2,4-D or of atrazine plus 2,4-D, each at 1.6 kg ha^{-1}, as a pre-emergence spray is common practice for cane grown in South and Southeast Asia, but metribuzin has qualified as an excellent grass killer in sugarcane plots. Other potent herbicides used in cane fields are velpar or hexazinone applied at 2 kg ha^{-1} as a pre- or post-emergence spray and these are effective in weed control. Glyphosate at 1 kg ha^{-1} is equally effective and is more readily decomposed in the soil than hexazinone. This author has compared Gesapax with other compounds like diuron in Burundi, Central Africa (Table 8.3).

Table 8.3 indicates that Gesapax at 4 or 8 l ha^{-1} and Karmex at 5 kg ha^{-1}, applied as a pre-emergence spray, was more effective in controlling weeds. Germination was not adversely affected by these herbicides. Gesapax-H was also tested and found to be more potent in the control of monocot weeds. Blackburn (1984) states that glyphosate and fusilade have been used to kill established cane stools and grass weeds prior to replanting.

Among the noxious weeds, Johnson grass (*Sorghum halepense*), a native of southern Europe and Asia, has reduced cane yield in Louisiana (USA), Fiji, Australia and India and it is difficult to control. However, 'Fenac' (2,3,6-trichlorophenyl acetic acid) applied at 2–3 kg ha^{-1} as a pre-emergence spray has achieved a fair measure of control of this grass. To control Burmuda grass (*Cynodon dactylon*) in autumn-planted cane in Taiwan, directed post-

Table 8.3. Effect of herbicides on germination and weed counts in planted cane (cv. NCO310) in Burundi, Central Africa. (Hunsigi unpubl.)

Treatment	Germination (%)	Weeds (number in 18 m^2)		Total weed count in 18 m^2	Total weed count $(X + 1)^{1/2}$
		Dicots	Monocots		
Gesapax (4 l ha^{-1}; 500 pw)	56.85	28.00	16.25	44.25	6.5250
Gesapax (8 l ha^{-1}; 500 pw)	60.25	22.75	19.50	42.25	6.5475
Karmex (diuron) (2.5 kg ha^{-1}; 80% wp)	58.97	31.00	47.00	78.00	8.3900
Karmex (5 kg ha^{-1}; 80% wp)	59.90	29.00	14.25	43.25	6.6295
Dalapon (2 kg ha^{-1})	49.97	33.75	31.50	65.25	8.0225
Dalapon (4 kg ha^{-1})	53.95	26.50	32.50	59.00	7.6900
Paraquat (1.5 l ha^{-1})	55.62	—	—	—	—
Unweeded check	54.60	42.50	40.75	83.25	9.1250
Weeded check	54.90	—	—	—	—
F test	NS	—	—	—	Significant
lsd 0.05	—	—	—	—	2.4103

[a] All herbicides were applied pre-emergence (3 days after planting), except paraquat which was used post-emergence (8 weeks after planting).
[b] Weed counts were not taken in paraquat as the herbicide was not sprayed; pw or wp, wettable powder.

emergence application (DPA) of 1.92 kg paraquat, plus 8.5 kg dalapon plus 1.65 kg 2,4-D ha^{-1} has been advocated and 81% control of this weed has been achieved (Peng 1984). In greenhouse tests, *Imperata cylindrica* was controlled by application of glyphosate (Round up) (loc. cit.).

Continued use of a single herbicide leads to a shift in the weed flora and results in a preponderance of one weed species. For example, if diuron is used, *Digitaria horizontalis* becomes a major problem because it is tolerant to diuron. Hence, a combination or 'cocktail' of synergistic herbicides is needed to control the broader spectrum of weed flora. The best examples of such combinations to control the broader spectrum of weeds in cane are diuron plus 2,4-D, atrazine plus 2,4-D (each at 2–2.5 kg ha^{-1}), TBA plus 2,4-D, and TBA plus MCPA.

The efficacy of herbicides is further improved when surfactants (surface active agents), such as emulsifiers, detergents and wetting agents, are used. Chemicals such as Citowett, Multifilm, Dalawet and even jaggery, molasses and common soap act as good surfactants. The threshold concentration of surfactants is within the range of 0.1–1.0% (Shoji 1965).

In many developing countries integrated weed management, i.e. a combination of manual, mechanical and chemical methods, is adopted to effectively control the weeds for the first 100 days in sugarcane. This is all the more necessary when sugarcane is intercropped with susceptible crops like groundnut, soyabeans, tomatoes and potatoes. Peng (1984) has proposed a 'selectivity index' especially for crop combinations and this is expressed as the ratio of ED20/ED80, where ED20 is 20% crop mortality and ED80 is 80% weed control. Thus, the selectivity index is the estimation of a compound's selectivity to a specific crop and an index of 2 or above is desirable for the crop combination. It has been shown that linuron and lasso have a selectivity index of 3 to 3.5 when cane is intercropped with susceptible crops like groundnut and soyabeans.

8.2 Pests

Sugarcane is known to be infested by 125 species of pests and, based on their feeding habits, they may be classified as (1) borers, (2) sucking pests, (3) subterranean pests, (4) defoliators and (5) non-insect pests (David and Nandagopal 1986). Moth borers are acknowledged as the most serious pests of sugarcane and some other important pests are: detailed here:

1. Early shoot borers (*Chilo* sp., Lepidoptera, Pyralidae, Crambinae). This pest is seen in Myanmar, China, India, Malaysia, Thailand and the islands of the Indian and Pacific oceans. *Chilo infuscatellus* Snellen is a serious pest in India, Java, Pakistan, the Philippines, Indonesia and Taiwan. Following the death of the leaf spindle the bored cane is called the 'dead heart' and tillers profusely, thus making up the plant population. Hence, a threshold percentage of 'dead hearts' is kept at 30 (Lakshmikantham 1983). Blackburn (1984) made similar observations and asserted that the joints bored, expressed as a percentage of all millable canes (PJI), should be 20 in order to warrant control measures. With serious infestations the crop stand is affected and there is a loss in yield and quality. Inundative releases of mass-multiplied egg parasitoids, *Trichogramma australicum*, *T. japanicum* and *T. fasciatum*, have proved efficient in controlling this pest.

2. Sugarcane borers (*Chilo auricilius* Dudgeon and *Diatraea saccharillis* Fabricius, Lepidoptera, Pyralidae, Crambinae). *Chilo auricilius* is distributed in the Old World and *Diatraea* in the New World. The former is known to cause serious damage in India, Indonesia, Malaya, Bangladesh and Pakistan while the latter is active in the USA (Louisiana), Mexico, Colombia, Venezuela, Peru, Brazil, Argentina and the Caribbean Islands. Typical symptoms are the presence of holes in the internodes with stunted growth, resulting in reduced yield and quality. The effective biological control agents for *Diatraea* sp. are the larval parasitoid *Lixophaga diatraea* Townsend and the egg parasitoid *Trichogramma fasciatum* Perkins.

3. Internode borer (*Chilo sacchariphagus indicus* Kapur, Lepidoptera, Pyralidae, Crambidae). This is an important pest found in India, Mauritius, Réunion and Madagascar. The larvae bore into the tender cane tops and dead hearts are rarely seen (David 1986). Its activity is drastically reduced in canes which have flowered. Among the highly resistant species and allied

genera of sugarcane are *Saccharum spontaneum* and *Erianthus ciliares* (David 1986).

4. Top borer (*Scirpophaga nivella* Fabricius or *S. excerptalis*, Lepidoptera, Pyralidae, Crambidae). It is found in Myanmar, China, India, Japan, Malaysia, Pakistan, Sri Lanka, the Philippines and Taiwan. The damage is due to mortality of the shoots. Cane growth is arrested and a typical symptom is 'bunchy top'. Biological control includes the release of *Trichogramma japonicum.*

5. Sugarcane stalk borer (*Eldana saccharina* Walker). This borer is indigenous to and restricted to Africa. The *Eldana* population increases with plant age, and ants (*Alta* sp.) are important predators which keep the pest under control.

6. Indian sugarcane leafhopper. The *Pyrilla* (*Pyrilla perpusilla* Walker, Homoptera, Lophopidae) occurs in India, Pakistan, Bangladesh, Myanmar, Sri Lanka and Indonesia. The pest sucks the sap from the under surface of the lower leaves causing yellowish white spots to appear. High levels of manuring, irrigation and lodging enhance Pyrilla build-up. The hoppers exude a sweet fluid (honeydew) on which fungus grows. The pest has been successfully managed in recent years by releasing the lepidopterous ectoparasite *Epiricania pyrillae* in northern India.

7. In many sugarcane-growing countries of the world, white scale insects (*Melanaspis glomerata* and *Aulacaspis tegalensis* Zehna) are becoming increasingly important. These are spreading fast in Asia and Africa and their infestation levels increase with drought and continuous ratooning. The entire cane is covered with scales, giving a greyish black colour. The cane becomes shrivelled with stunted growth and has a reduced internodal length, resulting in a lowered yield and quality of juice. Highly infected canes pose problems with milling.

8. Other insect pests of less economic consequence are the mealy bugs; these are ubiquitous and 35 species are recorded on sugarcane. But the pink mealy bug (*Saccharicoccus sacchari* Cockrell) is possibly responsible for 'yield decline' of commercial varieties. Mites (*Acarina, Archinida*) are of minor importance and about a dozen species infest sugarcane.

9. Whitefly (*Aleurolobus barodensis*) has a wide distribution. Drought and N deficiency accentuate the attack by whitefly. Spraying with 0.1% malathion 50EC or 0.2% endosulfan 35EC after detrashing is advocated to control Whitefly.

10. Frog hoppers (*Aeneolamia* sp. Homoptera, Cercopoidea). These spittle bugs occur in the New World in the tropics or subtropics and are mainly found in Mexico, Venezuela, Brazil and Trinidad. The adults feed on cane leaves causing narrow, brown necrotic strips at the site of feeding. The brown strips coalesce when the adult population is higher and the leaves are destroyed and the internodes stunted. Consequently, the yield and quality of cane is reduced. Most of the work has been carried out on *Aeneolamia varia saccharina* (Distant) in Trinidad and Frank Blackburn's contribution will be remembered for the control achieved in Trinidad. Ratoons are severely affected by frog hoppers. The control schedule consists of using three or four

insecticidal sprays. The first spray of carbaryl (85% wp) (at 2.8 kg in 28 l ha^{-1}) with sticker is used for control of the first brood and application is followed by two to three sprays of malathion (95% wp) at 1.4 l ha^{-1} to control the subsequent broods of frog hoppers. More recently, synthetic pyrethroids have been used successfully in Trinidad (Blackburn 1984).

11. White grubs (*Anomala* sp., *Holotrichia* sp., *Pentodon* sp., *Alissonotum* sp. and *Hetronychus* sp., Coleoptera, Scarapaeidae, Melolonthinae, Rutelinae and Dynastinae). These grubs are C-shaped and fleshy with a white or grey body. In some countries they are a sporadic pest, while in Australia and India they have threatened the sugar industry and are only second in importance to moth borers. The visible symptoms of attack include drying of leaves, drooping of inner spindles and, finally, wilting of the plant. The loss of cane yield is about 10–12 t ha^{-1} or even more if infestation is high. The clump is easily pulled out since the roots are damaged. Integrated pest management is the only solution controlling this pest. During the first summer showers in April–May (as in tropical India), the beetles emerge in large numbers to mate and these can be collected and killed. Collection of high numbers of beetles (*Holotrichia serrata*, Fabricius), and up to 300 000 in 1 day, has been achieved in Karnataka, India (Veeresh, 1977). Fluorescent-light traps can be used to attract the pest. An infested field may be ploughed and the grubs, pupae and beetles picked out by hand and destroyed. Wherever feasible, puddling the soil and using rice paddies for two or three seasons is an assured method of control. Varieties like M134/32, CO953 and BO3 are resistant to white grubs. Chemical methods of control include use of HCH dust (10%) (hexachlorocyclohexane, known previously as benzene hexa chloride, BHC; the γ-isomer of HCH is called lindane) at 10 kg active ingredient ha^{-1} or as an emulsion at 2 kg a.i. ha^{-1}, and the efficacy can be enhanced by use of farmyard manure which attracts the young grubs. Quinalphos G at 2–2.5 kg a.i. ha^{-1} reduces the grub population, increases the number of millable canes and yield. Successful biological control has been achieved in Hawaii and Mauritius (Blackburn 1984). The grubs of *Anomala orientalis* (Waterhouse) in Hawaii and of *Clemora smithi* (Arrow) in Mauritius were parasitized by *Campsomeris marginella modesta* (Smith) and *Tiphia parallela* (Smith), respectively.

8.2.1 Non-Insect Pests

Among vertebrate pests, rats and jackals frequently visit sugarcane fields, causing considerable damage and yield loss. Baiting with zinc phosphide in food grains or dry fish at a ratio of 1:19 controls these pests.

Blackburn (1984) prophesied that in the future the sugar industry will be plagued by the *Eldana* borer in Africa, *Chilo* sp. in the Old World and *Diatrae* sp. in the New World unless integrated pest control measures are adopted.

8.2.2 Integrated Pest Management

Environmentally friendly approaches should be employed to contain pest infestations and the application of pesticides should be need based. Build-up of natural predators or parasites should be encouraged and biological control measures be adopted wherever possible. Use of insect pheromones, chemosterilants and insect pathogens will go a long way towards controlling serious pests and diseases of cane. Some of the important components of integrated pest management in sugarcane are:

1. Planting healthy, hot-water-treated setts.
2. Observing clean cultivation practices.
3. Burning crop residues and trash in endemic areas.
4. Avoiding application of heavy N doses and applying N in split doses wherever feasible.
5. Avoiding waterlogging by providing adequate drainage.
6. Detrashing the cane or using self-detrashing varieties in areas where mealy bugs, mites, whitefly, scale insects and Pyrilla are a problem.
7. Avoiding moisture stress at critical stages like tillering and the grand growth phase.

Table 8.4. Insecticides recommended for the control of major sugarcane pests. (David and Nandagopal 1986)

Pest	Control measures
Early shoot borer (*Chilo infuscatellus*)	Soil application of gamma-HCH 20EC at 1 kg a.i. ha^{-1} at planting and, if needed, one more application at 45 days Whorl application of lindane 10 G at 0.9 kg a.i. ha^{-1}, 35 days after planting
Top borer	Carbofuran 3 G at 1 kg a.i. ha^{-1} or phorate 10 G at 3 kg a.i. ha^{-1} at the 3–5 months stage.
Stalk borer	Spraying monocrotophos 40 EC at 0.75 kg a.i. ha^{-1} after removal of dry leaves at the 5–6 months stage.
Eldana borer[a]	Sett treatment with dieldrin/aldrin by mixing 2 g wettable powder with water (1:1) and dipping for 15 min
Frog hoppers[a]	3% gamma HCH or 5% carbaryl dust; 3% malathion dust at 315 kg ha^{-1};[b] 2% carbophenothion (Trithion); or carbaryl (50% wp) (with sticker) at 2.8 kg in 28 l ha^{-1} and later sprays with malathion (95% wp) at 1.4 l ha^{-1} using an average of 3–4 sprays
White grubs	Soil application of HCH 10% dust at 10 kg a.i. ha^{-1} or quinalphos 5 G at 2.5 kg a.i. ha^{-1} mixed with an equal quantity of farmyard manure, applied immediately after first summer rain (April to May)

[a] After Blackburn (1984).
[b] Effective against gamma HCH resistant nymphs with little residual value.

8. Earthing, wrapping and propping should be used to prevent lodging (not possible under conditions of mechanized cultivation).

8.2.3 Chemical Control

Much progress has been made in the control of pests, starting from use of indigenous plant products, like neem (*Azadirachta indica*), tobacco leaf extracts, coal tar and oil emulsions, and inorganic compounds such as lead acetate and lime. Then followed organochlorine derivatives such as DDT, HCH, endrin and other compounds which were used as dusts or emulsions. Subsequently, organophosphate compounds were introduced, and the use of carbamates and combination of insecticides came into vogue during the 1970s (David 1986). At present, many systemic and contact insecticides, such as phorate, dimethoate, malathion, quinalphos, monocrotophos and carbamates like carbofuran, are in use. Great strides have been made in developing methods of application like foliar, whorl and soil application and seed treatment. Emphasis is placed on need based application rather than on blanket application of insecticides.

The insecticides recommended for the control of major pests of sugarcane are listed in Table 8.4.

8.3 Diseases

Sugarcane is known to be susceptible to about 100 diseases and more than 30 are of economic importance. These diseases can be classed as fungal, bacterial, mycoplasmal and viral and some of the important ones are discussed here.

8.3.1 Fungal Diseases

8.3.1.1 Red Rot (Colletotrichum falcatum)

This is a most serious disease and is aptly called the 'cancer of sugarcane'. It attacks all plant parts but the damage to stems and setts is of greatest concern. Typical symptoms are reddening of the internal tissues of the stalk with cross-wise white patches when the stalks are split longitudinally. Red rot of seed cane is widespread in Australia, Taiwan, Louisiana and northern parts of India. Waterlogged conditions accentuate this disease. Thermotherapy does not seem to control the disease. Agnihotri (1983) reported that fungicides like bavistin, benomyl, vitavax, aretan, topsin, etc. were effective in vitro but failed to control the disease under field conditions. The only remedy lies in the development of resistant varieties, and physiological resistance appears to be restricted to *Saccharum spontaneum*. Simple agronomic measures to contain the disease include the use of healthy setts, trash burning, crop rotation, provision of adequate drainage and restriction of the number of ratoons.

8.3.1.2 Smut (Ustilago scitaminea)

It was originally supposed that this disease belonged only to the Old World since *S. barberi* and *S. spontaneum* are highly susceptible, while *S. officinarum* is highly immune. However, it has been confirmed that the disease is spread throughout all countries lying between 20 °N and 20 °S of the Equator (Agnihotri 1983). Smut is seen in countries like Hawaii (USA), Mauritius, the Philippines, Taiwan, Cuba, Florida (USA), South Africa, Jamaica, Trinidad, Argentina, Zimbabwe, Sri Lanka, Brazil, Guyana, India, Pakistan and Bangladesh. The disease is easily diagnosed by the appearance of distinctive, black, whip-like structures covered by a thin white membrane; when ruptured a dense mass of chlamydospores is released. The disease is spread by wind, irrigation water and use of smutted setts. Yield loss is heavy: 50% or more. Primary infection can be controlled by heat therapy. An integrated approach to control smut consists of (1) selection of healthy seed material, (2) avoiding ratooning, (3) thermotherapy, (4) treatment of setts with fungicides, and (5) roguing and destruction of smutted clumps.

8.3.1.3 Wilt (Cephalosporium sacchari)

This disease is seen in many diverse sugarcane-growing countries like the Philippines, Uganda, South Africa, West Indies, Mexico, United States, India and many South American countries. The affected plants show stunted, non-uniform growth, and a typical symptom is the yellow midrib and green leaf lamina. The cane becomes 'tubular' (light and hollow) with the development of cavities in the stem and these contain abundant conidia. The transmission of wilt is through seed pieces and secondary infection through wind, rain and wilt affected soils. Moisture stress and a wide C:N ratio increase the incidence of wilt. Agnihotri (1983) claims to have controlled the disease with water treatment in the presence of fungicides (0.1%) like benomyl, bavistin, thiram and aretan. This author believes that it is necessary to carefully remove and destroy affected plants. Development of wilt-resistant varieties is the sure way to contain the disease. Application of Mn/B to the soil or to healthy setts soaked in 40 ppm of these nutrients in solution offers a fair degree of control.

8.3.1.4 Eye Spot (Helminthosporium sacchari)

Among the leaf spots caused by fungi, eye spot is the most serious since, under severe conditions, a drastic reduction in the yield and quality of cane is observed. This disease, also called *Helminthosporiose*, occurs in 66 sugarcane-growing countries, including Hawaii, Puerto Rico, Cuba, Taiwan and Florida. In South India it was found to be associated with low pH, low soil potash and high levels of N. The incidence was more severe in November-December when the nights are cooler and humidity is high (due to the surrounding flooded rice cultivation). Cloudy weather accentuates the disease. As the temperature rises during February the incidence is reduced. The symptoms are round or oval, elongated spots on the leaves; these have a reddish centre and brown border. After about a week, the reddish brown streaks or 'runners' develop towards the leaf tips along the veins. The severity of the attack is due to the toxin Helminthosporioside, which reduces the Fe content of the leaf and impairs chlorophyll synthesis. This disease can be controlled with two or three sprays of 0.2% copper oxychloride at 15- to 20-day intervals just before the winter season. Development of a resistant variety is the final answer. Among the commercial varieties grown in Mandya (India) CO62175 was less susceptible than CO419.

Other leaf spots such as that caused by *Cercospora* sp. (yellow spots), are not that serious and can be easily controlled by a spray of 0.2% copper oxychloride.

8.3.1.5 Pineapple Disease (Ceratocystis paradoxa)

This disease has spread far and wide in Brazil, Hawaii, Mauritius, South Africa, Taiwan, Puerto Rico, Australia, India, etc. The affected cane smells like mature pineapple. Major transmission of the pathogen takes place through the seed pieces (cut ends). Dipping of setts, or dipping of sickles or sett-cutting tools in phenyl mercury acetate (PMA), benomyl (0.1% solution) is preferred to use of agallol or aretan. Sett treatment has improved germination due to some hormonal or synergestic effect.

8.3.1.6 'Pokkah boeng' (Gibberella fujikuroi)

It is worldwide in distribution. The disease 'pokkah boeng' is named after a Japanese term meaning malformed or twisted top; POJ2878 is highly susceptible. The visual symptoms include narrow, short leaves which are chlorotic at the base. At an advanced stage top rot takes place along with death of the cane. This disease can be best controlled by growing resistant varieties.

8.3.2 Bacterial Diseases

8.3.2.1 Ratoon Stunting Disease

For a long time RSD remained an enigma and was thought to be of viral origin, but recent results have demonstrated that RSD is caused by a coryneform bacterium. The general symptoms are reduced tillering, a poor root system, stunted growth, thin stalks with shorter internodes and yellowish foliage. The typical symptoms of RSD can be observed by longitudinally splitting mature cane, where upon orange-red or yellow-orange, usually pink-red and reddish brown discoloration can be seen at the nodal regions. RSD is responsible for 'run-out' of many varieties. It is transmitted through setts taken from diseased plants, or by infected cane knives and cutting blades. Hot water or a moist hot air treatment controls RSD. The insect vector for RSD has not yet been identified (Agnihotri 1983). The development of tolerant varieties would give a lasting solution and H60-6909 and B41242 are found to be highly resistant to RSD (Agnihotri 1983).

8.3.2.2 Leaf Scald (Xanthomonas albilineans)

This is the most 'trouble-making' disease and is found in all cane-growing countries. The characteristic symptoms are white stripes running along the full length of the leaf blade and sheath, and etiolated leaves. The drying out of leaves takes place from the tip downwards and gives a scalded appearance. The internodes are stunted with abundant side shoots (*lalas*). Juice quality is greatly impaired. Leaf scald is transmitted through infected setts, cane knives and reaping machines. Resistant clones can be observed in *S. spontaneum*. Thermotherapy, use of healthy seed material and resistant varieties and high soil fertility coupled with adequate internal drainage will reduce the incidence of this disease.

8.3.2.3 Gummosis (Xanthomonas campetris pv. vasculorum)

This is a minor disease except in Puerto Rico, Mauritius, Brazil and Australia where it causes extensive damage. Apart from reducing the yield by causing the death of foliage, the presence of gum impairs the crystallization of sugar, and leads to lowered sugar recovery, and adversely affects milling operations. The external symptoms are yellowish to orange stripes on mature leaves and the tissues slowly dry up. The exudation of yellow gum from the cut ends and basal portion of the infected stems gives the name gummosis. The infection is transmitted by agricultural implements and diseased setts. Use of disease-free canes and resistant genotypes helps to check the disease.

8.3.3 Mycoplasmal Diseases

8.3.3.1 Grassy Shoot Disease (GSD)

This is prevalent in India, Myanmar, Sri Lanka and Sudan and is similar to white leaf disease occurring in Taiwan and Thailand (Agnihotri 1983). The external symptoms are profuse tillering with narrow leaves and a varying degree of chlorophyll loss. The clump appears like a grass, hence the name grassy shoot. One or two millable canes from the affected clump appear apparently healthy but when planted show distinct symptoms of GSD. The disease is caused by a mycoplasma-like organism (MLO) and transmitted by insect vectors like *Melanaphis sacchari*, *M. indosacchari* and *Aphis maidis*. Ratoons manifest the disease more than the plant crop. The albinoid leaves contain less Mn and, hence, GSD-affected plants require more Mn. In general, GSD-affected plants contain higher amounts of amino acids, amides and other organic acids. GSD can be controlled by thermotherapy and strict roguing.

8.3.4 Viral Diseases

8.3.4.1 Sugarcane Mosaic Virus

This viral disease is widely distributed in sugarcane throughout the world. It is transmitted by setts and aphid vectors. The external symptoms are seen more on the basal portion of younger leaves than of old leaves. The chlorotic yellowish stripes alternate with normal green areas on leaves, thus giving a mosaic appearance. Chemotherapy or thermotherapy are of little success in the control of mosaic and resistant varieties are the sure solution for this malady. *S. officinarum* is highly susceptible to mosaic, whilst *S. spontaneum* and *S. sinense* are highly resistant (Blackburn 1984) and their inclusion in the breeding programme will control the disease.

8.3.4.2 Sugarcane Fiji Disease

Fiji disease, caused by a virus, is found in the islands of the South Pacific, Australia and Thailand. It was first studied in Fiji and hence, acquired its name. A typical symptom of the disease is the occurrence of galls on the underside of leaves. Near complete control has been achieved by use of resistant varieties.

8.3.5 Nematodes

Nematodes are known to be parasitic and about 50 species are associated with sugarcane. Infestations seem to be more serious in light soils. Broadly speaking, there are ecto and endoparasitic nematodes. General chlorosis, stunting and wilting are the above-ground symptoms, while root galls, root lesions, stubby roots and a necrotic root surface are the below-ground symptoms. Nematicides like D-D, telone, EDB and DBCP control these pests. Among the new nematicides, Vapam and Nemagon, Vapam is more effective as it can be easily drenched. Proper crop rotations are also effective in controlling nematodes.

Chapter 9 Ripening and Harvest

9.1 Ripening

According to Clements (1959), ripening is a physiological senescence which occurs between rapid growth and ultimate death of the plant. In essence, ripening is a natural and parallel function of maturity and the latter is the culmination of a diphasic physiological process. In the first phase, the ageing leaf detaches from the subtending internode and in the second phase sugar accumulates in the fully expanded internode. The maturation process is not divorced from varietal, cultural and ecological factors. The second phase of maturity is controlled by N levels and tissue moisture and indexes have been developed to monitor these features. Planned exhaustion of tissue N and H_2O are the primacy for achieving ripened cane. Alexander (1973) described this most succinctly by saying that, 'the plant water regime most effective in promoting cane ripening is that which gives the greatest restriction against growth while still maintaining a normal source to sink system for sugar synthesis, transport and storage'. As the cane is ripened, so the sheath moisture should drop from about 85 to 72% and the N index should fall from 2 to 1.25%.

Cane is said to be ripened when growth has practically ceased and elongation of the stalks is arrested or minimized with the concomitant accumulation of sugar. The plants' concern is sucrose retention rather than overutilization. Environmental factors like high temperature, humidity, soil moisture and tissue N are growth promoting and are not conducive to sucrose accumulation. As mentioned above, flushing out of tissue N and H_2O are a prime requisite for cane ripening. In unripe cane, the activity of acid invertase is high.

In the early stages of growth, when the plant is actively growing, tissues have high levels of N, moisture and enzymes like invertase and phosphatase, more invert sugars and intermediate products of photosynthesis and increased N metabolism with a high respiration rate. With ageing, and where ripening conditions exist, there is a gradual exhaustion of N, water, lowered reducing sugars and activities of related enzymes with a concomitant build-up of sucrose. It has been observed that countries which lie between 18 °N and 18 °S latitude have, by and large, better ripening conditions and tend to accumulate sugars. It has also been surmised that cane maturity is a function of age, N and moisture status but is complicated by climatic parameters like light, temperature, humidity, rainfall, etc.

From the foregoing, it has been discerned that the plant must be 'shocked' before it turns from the vegetative to the sugar stage (Humbert 1976). Several methods, ranging from cultural techniques and use of oils, growth promoters or inhibitors to use of nutrients, defoliants, desiccants and ripeners, have been employed to ripen cane (Table 9.1). Withdrawing water 4–6 weeks prior to harvest ('cut-out' period) has been successfully used in non-monsoon climates where rains do not interfere in the natural ripening process. Wrapping and propping the standing cane to prevent lodging was a time-tested ripening method in India. Defoliants, desiccants, antibiotic and antimicrobial agents have also been tested as ripening agents and results have met with variable success.

Table 9.1. Methods used to Ripen Cane

Cultural	Withholding irrigation; tying or supporting (wrapping and propping) the cane
Plant/animal products	Molasses, petroleum products such as Sucrol (ESSO 59 G), FS 40 (diesel oil plus pentachlorophenol plus 2,4-D)
Growth promoters	2,4-D, GA, 2,3,6-trichlorobenzoic acid (TBA)
Growth inhibitors or antiauxins	Malic hydrazide, Cycocel, abscisic acid (ABA)
Plant nutrients	Boron, mono-potassium phosphate, molybdenum, manganese
Defoliants or desiccants	Compounds like 15302, 15303
Antibiotics or antimicrobial agents	Naramycin A, streptomycin sulphate, magnamycin, nystatin, novobiocin
Chemical ripeners or sucrose enhancers	Cycocel (CCC), 2-chloroethyl trimethyl ammonium chloride Polaris (glyphosine; code CP 41845) N, N-bis(phosphonomethyl)glycine Ethrel or Ethephon, 2-chloroethyl phosphonic acid Embark (mefluidide), N- [2,4 dimethyl-5-(trifluoromethyl) sulphonyl) amino)phenyl] acetamide Fusilade super (code PP005) or fluazifopbutyl, [butyl-2-[4-(5-trifluoromethyl-2-pyridinyloxy) phenoxy] propanoate Dowpon (dalapon), 2,2-dichloropropionic acid Asulox(asulam), methyl sulfanil-Yl-carbamate Polado (glyphosate; code Mon 8000), sodium sesquisalt of N-(phosphonomethyl) glycine Round up (glyphosate; code MOM 2139), iso-propylamine salt of N-(phosphonomethyl) glycine

Growth promoters, like 2,4-D, were first tested in Cuba in 1949 and subsequently many inhibitors, including malic hydrazide (MH), TIBA, ABA and CCC, were tested but the response was not consistent. However, among plant nutrients, B, Mo and KH_2PO_4 improved ripening by converting reducing sugars to sucrose in the top internodes of the cane (Table 9.2). Alexander (1973) argued that some of these plant nutrients controlled enzyme activity; he found that Mo inhibited phosphatase activity, while Si inhibited invertase activity.

The advent of ripeners or sucrose enhancers heralded a new era in cane culture. Alexander (1973) found that 6-azauracil decreased the activity of invertase and co-administration with GA or Si enhanced the ripening effect. Time-tested ripeners include Polaris, sodium or calcium metasilicate, Embark, ethrel and asulam, all of which act by ripening immature cane and enhance the sucrose content. It is anticipated that for short periods these chemicals can arrest vigorous growth and the accumulated photosynthate can be partitioned between sucrose and non-sucrose. The ripeners are recognized to control the activity of invertase; therefore, the suggestion has been mooted that the ratio of acid invertase (AI) to neutral invertase (NI) is a good index of the ripening process, as shown in Table 9.3

Table 9.2. Effect on pol and reducing sugars in the top internodes of cane at 10 and 20 days after chemical spraying (cv. CO62175). (Hunsigi and Suryanarayana unpubl.)

	pol (%)		Reducing sugars (%)	
	10 days	20 days	10 days	20 days
Control (water spray)	10.87	12.89	1.25	1.03
Polaris (1 kg ha^{-1})	11.99	13.56	1.73	0.95
Polaris (3 kg ha^{-1})	11.60	11.27	1.70	0.88
Sodium metasilicate (1 kg ha^{-1})	13.50	13.28	1.70	0.98
Sodium metasilicate (3 kg ha^{-1})	14.64	14.21	1.18	0.77
KH_2PO_4 (10 kg K_2O ha^{-1})	12.41	12.59	1.28	0.95
KH_2PO_4 (20 kg K_2O ha^{-1})	14.29	13.24	1.85	0.77
Boron (50 ppm)	12.36	14.34	1.40	0.98
Boron (100 ppm)	11.61	15.38	1.55	0.77
Molybdenum (20 ppm)	12.60	14.04	1.35	1.00
Molybdenum (40 ppm)	13.44	13.08	1.25	1.03

Table 9.3. The ratio of AI/NI with use of different ripening sprays. (Hunsigi and Suryanarayana unpubl.)

	Days after spraying ripener				
	0	1	2	5	7
Control (water spray)	1.23	1.03	0.95	1.30	1.15
Polaris (0.4%)	1.21	0.36	0.86	0.89	1.28
Sodium metasilicate (0.5%)	1.28	0.67	0.57	0.76	1.11

Obviously a reduced ratio of AI/NI suggests enhanced ripening with a consequent increase in sucrose. The effect of ripeners lasts for 6–8 weeks and is conditioned by temperature, moisture, rainfall and cane variety. Among the varieties, high tonnage ones like CO62175 in India, B51129 in Jamaica and NCO376 in South Africa are more responsive to ripeners as they convert invert sugars in the top internodes to recoverable sugars. The findings of Clowes (1980) are in harmony with these results and he obtained a good response with vigorous varieties like NCO376, NCO310, NSS/805, NCO293 and N52/219.

Among the ripeners, Polaris and CCC were first tested in 1969, followed by ethrel (Ethephon). Polaris has been used commercially in Brazil, Guyana, Florida, Hawaii, India, Jamaica, Mauritius, Louisiana, Mexico, Malawi, Swaziland, Panama and South Africa. Essentially, ethrel and Polaris enhance sucrose in the immature portion of cane (i.e. top internodes). Ethrel was successfully used on NCO376 in South Africa at a rate of 1 to 1.5 kg ha^{-1} and the response ranged from 0.8 to 1.4 t ers ha^{-1} (Donaldson and Van Staden 1989). The term ers represents estimated recoverable sugar which can be expressed as: (sucrose − 0.485 [non-sucrose%] − 0.056 [fibre %]). A distinct advantage with ethrel seems to be the increased cell volume and size, and additional sucrose can be stored even if ripening has not occurred. More important is that ethrel has no adverse effects on succeeding crops and it controls flowering with reduced pithiness and improved cane quality. Embark may be used on mature cane stalks and is a possible alternative to ethrel, although it has not been extensively field tested.

Hilton et al. (1980) in Hawaii, Clowes (1980) in South Africa and McCatty (1980) in Jamaica have demonstrated that glyphosate salt, designated Mon 8000, at 0.25–1.00 kg a.i. ha^{-1} acts for a longer time (6–8 weeks) and sucrose improvement can be observed under both rain-fed and irrigated conditions. Table 9.4 clearly indicates that Mon 8000 at 0.3 and 0.6 kg ha^{-1} and Polaris at 4.5 kg ha^{-1} are superior to the rest of the ripeners. Hilton et al. (1980) showed that the mode of action of Mon 8000 is similar to that of Polaris in

Table 9.4. Response of cane cultivars to ripeners as pol % cane. (Hilton et al. 1980)

Cultivar	Ripener (kg ha^{-1})					
	Mon (0.3)	8000 (0.6)	Embark (1)	Ethrel (1)	Polaris (4.5)	Control —
H59-3775	14.29	15.43	11.19	10.33	13.67	10.95
H57-5174	11.34	11.62	10.50	10.91	10.60	10.71
H54-775	11.05	12.25	9.85	10.08	11.68	10.99
H61-467	11.44	13.79	11.39	9.87	10.63	9.51
H56-4848	10.02	11.53	9.87	9.74	9.83	9.60
H60-7490	8.12	10.19	8.52	8.92	9.30	8.87
H56-278	13.55	14.83	13.02	12.78	13.29	12.82
Mean	11.40	12.80	10.62	10.37	11.28	10.49

that it retards apical growth while sugar synthesis, transport and storage continue. The most important feature of Mon 8000 is that it is more consistent than Polaris, with fewer cultivar differences, less sensitivity to excess N and no adverse effects on ratoons. With the radio tracer technique, Hilton et al. (1980) observed that ^{14}C could be traced more in sucrose. There was a lower respiratory rate, lowered activity of invertase and inhibition of protein synthesis.

Ripeners like ethrel and Polaris increase the sucrose content of immature stalks, but compounds of glyphosate, viz. Polado and Round up (0.67 and 1.0 kg ha^{-1}, respectively) improve the sucrose content of mature stalks and are aptly called sucrose 'loaders'. These loaders have a greater potential than the ripeners but an overdose should be avoided since they cause poor ratooning. The average response of these loaders is 3.2 Mg sucrose ha^{-1} (Donaldson and Van Staden 1989). Further confirmation comes from Clowes (1980) who observed that glyphosates increase the sugar yield of mature, actively growing cane by 'loading' sucrose in the lower and more mature cane stalks. A mean increase of 1.5 units of sucrose has been achieved with glyphosate.

A new product, coded as PP009 and known as Fusilade, showed an ability to ripen cane. A more concentrated formulation containing Fluazifopbutyl (125 g a.i. ha^{-1}) and named Fusilade Super showed considerable promise at 300–400 ml product ha^{-1}. An average increase in sugar is 1.5 Mg ha^{-1}. The effect of Fusilade Super is more consistent than that of Polado and ethrel and lasts for 4–9 weeks (Donaldson and Van Staden 1989). Also, the low cost of Fusilade Super and its minimal residual activity are factors which favour its wider acceptance. The observations of Soopramanien et al. (1990) in Mauritius further established the superiority of Fusilade Super over Polado (glyphosate) and ethrel (Ethephon) in improving cane quality.

Ripening by use of these chemicals is achieved through restriction or reduced growth and is accompanied by symptoms such as:

1. Bushy plants with excessive tillers.
2. Side shooting due to loss of apical dominance.
3. Shortening of upper internodes.
4. Reddening of leaf sheaths.
5. Development of brown or yellow tinges to the leaf canopy.
6. Death of spindle leaves and apical region (typical symptom from use of Fusilade Super).
7. Necrotic rings on the internodes.
8. Suppression or reduced flowering.

   Care should be taken that the death of the spindle or a reduced leaf canopy does not result in a yield loss which may negate the gains from improved cane quality.

Donaldson and Van Staden (1989) have listed additives or adjuvants, i.e. KCl, Agral 90, Penetrex, etc., to be mixed with standard ripeners to improve their ripening effect.

9.2 Harvesting

Harvesting is carried out when the cane has ripened, which ensures that there is a peak sugar content. Blackburn (1984) showed that when the sucrose content was plotted against latitude there was a bimodal curve with its nadir at the Equator and peaks at 18 °N and 18 °S. In general, sugar accumulation is high during the cooler months and this also coincides with the harvest period. In irrigated areas in Peru and Hawaii the cane is crushed throughout the year, while in Brazil, Argentina, Fuji, Mauritius and South Africa harvest may begin in May and end by October. The harvest in the Philippines is for 10–12 months, in Louisiana it is for 3–4 months (October–January), while in Florida and India it is for 6 months (October–April). Puerto Rico and the Caribbean islands also crush cane for 6 months (January–June). It is envisaged that the varietal spectrum of early, mid and late varieties should be used which will not only improve overall sugar recovery but will also increase the grinding season.

Pre-harvest surveys are made and cane harvesting is regulated. In some well-developed factories, representative cane samples are taken and detailed analyses for brix, pol and purity are made in the laboratory. The common practice is to test the standing cane with a hand refractometer.

The term 'harvesting' includes cutting, loading and transport. In most countries of Asia, Africa and South America, cane is cut manually which gives a much cleaner cane. Top removal is of prime importance as top internodes are immature (unless ripeners have been used). Hand stripping reduces the trash to an acceptable range of 2–3%. In many developing countries cane is transported by animal-drawn carts, but, increasingly, tractor-trailers (4–5 t capacity) and trucks of 10–20 t capacity are used for cane haulage. The Bell self-loading trailer (2–6 t capacity) developed in South Africa is used in Trinidad and the Philippines.

The Jacobyl trailer is extremely versatile and can be rear loaded or side loaded. Sometimes even a chain of trailers is used for cane transport. The Mascane system is an example of this, where four trailers travel in tandem behind one tractor and it is economical for longer distances. This system is widely used in Kenya, Malawi, South Africa, Tanzania and Zambia (Blackburn 1984).

Mechanized harvesting has become mandatory, due to the non-availability and/or high costs of labour, in Louisiana, Florida, Hawaii, Puerto Rico, Australia and South Africa. Australia maintains a traditional clean cane supply to the factory.

In South Africa the trash is conserved and green-cane harvesting is carried out. When mechanized harvesting is introduced, it must be integrated into the total system from planting through cultivation, harvesting and transport to milling. One of the earliest contrivance after the Second World War was the push rake and grab system of harvesting. In general, there are two types of mechanical harvesters, viz. whole stick harvesters and chopper harvesters.

The former are used in a 12-month crop, where the stalks are straight, about 3 m tall and yield 100 $Mg\,ha^{-1}$. They require the service of mechanical loaders and maintain a daily output and a reasonable standard of topping. Whole stick harvesters generally gather, top and cut the cane before depositing it in bundles along and across the row. The cane is then loaded into a tram truck or on to motor transport by a front-end and grab-type loader. Overall cleanliness depends on effective pre-harvest burn and the skill of harvesters and loaders.

Chopper harvesters are employed for a long crop cycle (24–36 months), highly tangled and severely lodged cane. These are complete and neat harvesters, which operate on a gather, top, cut, billet and load principle (Gaunt 1965). They discharge the harvested and chopped material into containers or bins.

Among the whole stick harvesters, the soldier harvester (Cameco) operates on green cane and topping is effective in erect or semi-lodged fields. Gathering arms and chains lift the partially lodged cane while the base is cut by the adjustable bottom cutter blades. The tops are removed by a topping device (Irvine 1977). The whole stalks are heaped and fired to remove the leafy trash, and grab loaders place it in a wagon or trailer ready for transport to the mill or transloading station.

According to Irvine (1977), over the years mechanical harvesting has evolved from grab harvesters to push rake and 'v' cutters. There is very little damage done to the stools by the 'v' cutters and a pick-up buggy was developed by the Hawaiian Sugarcane Planters' Association (HSPA) to deliver the cane to the roadside. Blackburn (1984) points out that the push rake system is a widely adopted method of reaping irrigated plantations despite its major defects, like uprooting the stool and incorporating large amounts of extraneous matter (including rocks, twigs etc.). Consequently, the cane has to be 'laundered' and a special cleaning device is used before processing; this involves floating the harvested material across a water gap in which mud, rock and heavy objects are removed.

In Hawaii, a 2- to 3-year crop cycle is used and the yields exceed 300 $Mg\,ha^{-1}$ with a tangled mass of vegetation. The mechanical harvest of Hawaii's recumbent cane, virtually a forest of tangled stalks, poses a serious problem.

The chopper harvesters were developed in Australia and the Australian cane combine is the most widely used system. The combine harvester pushes the stalks, cuts them at the base into pieces or billets and an air blast removes the trash. The partially cleaned and chopped canc is then dropped into a tractor-drawn bin travelling by the side of the combine. Mechanized harvesting is carried out after burning the cane. Use of desiccants like gramoxone (1.5–3.0 $l\,ha^{-1}$) is suggested. But, more recently, green cane harvesting has been carried out by modified combine harvesters. The Cary cane combine harvesters used in the Caribbean are described by Deacon (1968). The fallen canes are raised and cut at the base by short, serrated knives. A rotating two-bladed chopper knife cuts the cane into small pieces and an air blast removes the dirt and trash.

Fig. 9.1. Tractor-mounted sugarcane harvester (Sharma and Singh 1988)

The chief defect of the chopper harvester system is the high infestation with *Leuconostoc*, due to a higher number of cut ends. *Leuconostoc* souring can be controlled by treating the knives, harvesters, etc. with a suitable bactericide.

In conclusion, it can be stated that the mechanization of cane culture from planting to harvesting in the developing countries of Asia, Africa and Latin America would possibly lead to socio-economic problems with disastrous consequences. Hence, a partial mechanization seems to be the answer. Sharma and Singh (1985) have pleaded for partial mechanization and have developed the IISR tractor-mounted sugarcane harvester which serves the purpose of basal cutting of a single row of sugarcane stalks and wind rowing of the harvested crop (Fig. 9.1). Humbert (1968) aptly stated that 'mechanization must of necessity proceed slowly to not create more problems than it solves'.

Part II

Chapter 10 Fibre and Energy Cane

10.1 Fibre Cane

With more than five decades of commercial exploitation, sugarcane could possibly be renamed as alcohol cane or energy cane and a futuristic term could be fibre cane, since bagasse may become the most important source of agricultural fibre for the pulp and paper industry. Bagasse or megass is the fibrous residue from cane following crushing in the mill. By the word fibre one means all the insoluble solids – fibrous or otherwise (Meade and Chen 1977).

The heterogenous character of bagasse fibre needs to be reckoned. The mill wet bagasse has a 50% moisture content and is used as a captive fuel in the sugar industry. It is prophesied that bagasse fibre will surge ahead of all other non-wood fibres and will become the most important raw material for the paper industry.

Recent FAO (1988) estimates show that global production of sugarcane is 988 million tons, and bagasse production is approximated at 247 million

Table 10.1. Area and production of sugarcane bagasse and pulp in different continents and regions of the world. (FAO 1988)

Continent/region	Sugarcane area (10^6 ha)	Cane production (10^6 t)	Approx. production of bagasse[a] (10^6 t)	Approx. bagasse available for pulp[b] (10^6 t)	Approx. production of bagasse pulp[c] (10^6 t)
World	16.349	988	247.00	12.35	1.764
Africa	1.197	73	18.25	0.912	0.130
Northern central America	2.790	181	45.25	2.262	0.323
South America	5.206	327	81.75	4.087	0.584
Asia	6.750	374	93.50	4.675	0.668
Oceania	0.403	31	7.75	0.387	0.005
Developed all	0.975	78	19.50	0.975	0.139
Developing all	15.374	910	227.5	11.375	1.625

[a] About 25% is approximated as the production of mill wet bagasse from sugarcane.
[b] A minimum saving of 5% bagasse is envisaged for paper pulp.
[c] 7 t mill wet bagasse (50% moisture) would yield approximately 1 t paper pulp.
Production figures in metric tons.

tons. Assuming a 5% saving in bagasse for paper pulp, about 1.76 million tons of bagasse pulp is available for paper making (Table 10.1). Atchison (1986) has indicated that the world production of bagasse pulp is expected to reach a level of 2 million tons. In economic terms, it will continue to lead all other plant fibres and may represent 2–3% or more of the world pulp production (Atchison 1976).

10.1.1 Fibre Accumulation

Fibre accumulation starts from about 3 months and continues to increase with age. The millable cane (trash free) is made up of approximately 65.75% water and the remaining percentage is divided between fibre and soluble solids (Fig. 10.1). The quantities of these components are genetically controlled. The noble varieties of sugarcane are richer in water and sugar but relatively low in fibre. On the other hand, the interspecific hybrids are higher in both fibre and sugar. The fibre content of different parts of the cane stalk varies considerably. The bottom portion, called the 'butt', contains the most fibre and the nodal portion contains at least 2% more fibre than the internode (Narasimhan 1957).

Physiologically, sugarcane is tailored to produce high fibre. But the lower limit of fibre for maintenance of the plant structure is 5% while the highest possible sugar accumulation is 27% with a minimum of 70% moisture (Bull and Glasziou 1963). Nevertheless, the fibre content of cane is greatly controlled by the activity of the invertase enzyme. Alexander (1973) has suggested that the increased activity of acid invertase (AI) in *S. spontaneum* favours fibre production, while in *S. officinarum* increased activity of neutral

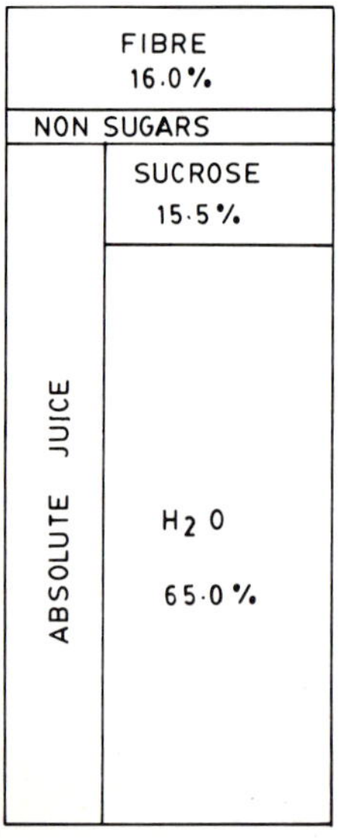

Fig. 10.1. The composition of well-ripened sugarcane

invertase (NI) favours sugar accumulation. Hence, the ratio AI/NI could possibly be used to decide fibre or sugar accumulation. A shift towards a higher AI/NI ratio tends towards fibre accumulation, while a lowered AI/NI ratio tends towards sugar accumulation. However, it is admitted that the thesis of an AI/NI ratio controlling fibre or sugar accumulation is an oversimplification of a complex phenomenon. Stevenson (1965) has given a genetic explanation for fibre production in cane. According to him there are mutant genes, originating as mutant alleles, in *S. spontaneum* which favour photosynthetic conversion to fibre, whereas in *S. officinarum* there are a few genes located at different loci which are responsible for the inheritance of a high sucrose trait.

10.1.2 Stem Anatomy in Relation to Fibre

The sugarcane stem is a meristele, i.e. vascular bundles are scattered throughout the cross section. A study of its anatomy reveals the epidermis, narrow cortex or rind and vascular bundles scattered in the ground tissue (Fig. 10.2). The epidermis is the outermost layer and has a fine coating of wax in the form of rods (Fig. 10.3). It has two types of cells, namely, long rectangular cells and short cells. The latter can be either cork or silica cells which give strength to the stem and toughness to the rind.

The cork cells are suberized and thin walled. As reported by Van Dillewijn (1952) the silica content varies from 0.4% in young leaves to more than 6% in old ones. The average silica content per ton of millable cane has been reported as 2.27 kg (loc. cit.) and the highest silica content is found in the trash. Silica has an adverse effect on pulp and paper making. Immediately below the epidermis is the sclerenchymatous hypodermis. Typically, fibro-vascular bundles of cane stem have a sclerenchymatous sheath which encloses both the xylem and phloem. The vascular bundles in the stalk

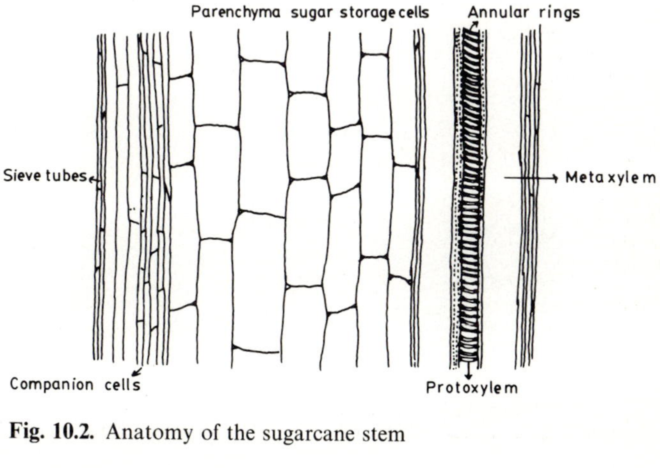

Fig. 10.2. Anatomy of the sugarcane stem

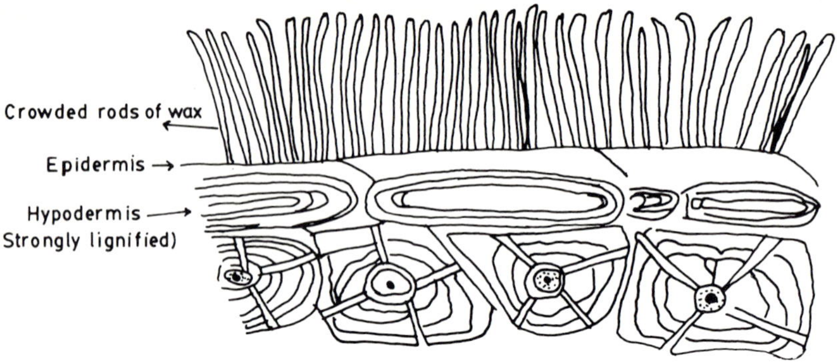

Fig. 10.3. A section through the sugarcane stem showing the waxy coating. (van Dillewijn 1952)

account for the major portion of fibre in bagasse. As the cane age increases, so there is deposition of lignin-like compounds in and around the fibrovascular bundles and this results in a hardening of fibres. The xylem or wood mainly consists of protoxylem, while the phloem or bast is made up of sieve tubes and companion cells. Parenchyma cells form the sugar storage cells (Fig. 10.4). As stated earlier, the vascular bundles are embedded in a matrix made up of parenchyma cells. In the growth ring, sclerenchyma is replaced by unlignified collenchyma cells. The protoxylem is also covered by a sclerenchymatous cap. It should be emphasized that the high proportion of sclerenchyma and collenchyma, in addition to wood and bast fibres, imparts the desirable tensile fibre properties for making paper of all grades.

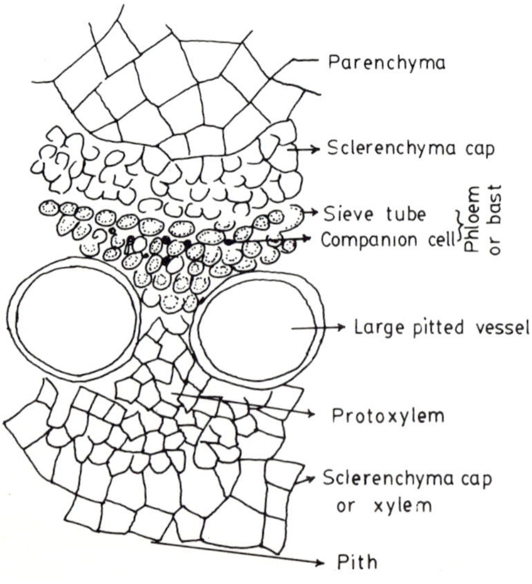

Fig. 10.4. A cross section through the sugarcane vascular bundles

10.1.3 Tissue Moisture vs Fibre and Sugar

A highly significant correlation exists between moisture and fibre and moisture and sugar. Thus, a large growing body of evidence indicates that 'tissue moisture' is like 'excess baggage', which can be thrown out by the plant to improve both sugar and fibre yields. Clements (1980) believes that the sheath moisture, an index of tissue moisture, can be kept as low as 68–69%. The recent technique of ripening involves the continued build-up of dry matter (half of which is sucrose and the other half fibre), and the biological pathway should be chosen so as to increase sucrose and ligno-cellulose at the expense of tissue moisture but without killing the plant.

10.1.4 Some Factors Influencing Fibre in Cane

There is tacit evidence to show that fibre content increases with drought conditions. However, this is accompanied by an increased pith content and a consequent reduction in useful fibres. It has been observed that drought-resistant varieties have well-developed mechanical tissues surrounding the vascular bundles and the sclerenchyma cells forming these tissues are more lignified. A lodged cane has less fibre. It is reasonable to assume that the lodged cane is more succulent with a higher tissue moisture and, hence, reduced fibre and sugar contents. Among cultural practices, row spacing has a profound influence on biomass and fibre yields. A narrower spacing of 0.5 m tends to yield more biomass and fibre than a wider spacing of 1.0 or 1.5 m.

One interesting study by Buren (1976) suggested that spraying a cane ripener like Polaris improved both pol and fibre content by depressing tissue moisture (Table 10.2).

Among plant nutrients, N has a significant influence on fibre content and increasing levels reduce the fibre percentage in cane. It has been pointed out that tissue moisture and N in shoots are intimately related and a strong positive correlation is observed at the early stages of crop growth. Excessive N with adequate water and low radiation causes succulent cane stalks. Large, thin-walled cells can neither be filled with sugar nor develop into fibres. In essence, a balanced carbohydrate – N relation is essential to accumulate both sugar and fibre. Eventually, following high N fertilization

Table 10.2. Effect of Polaris at 4 kg a.i. ha^{-1} on the stalk component of H59-3775 at Mahna Key Sugar Co. (Buren 1976)

Parameter	Check	Treated	Change
Soluble salts (%)	13.3	16.8	25.9
Fibre (%)	12.4	12.8	3.3
Moisture (%)	76.1	72.6	– 4.5

coupled with adequate soil moisture, thin-walled cells develop 'pithiness' or 'piping'.

It is reasonable to regard that a direct relation exists between sugar, fibre and moisture and maximum sugar accumulation should not be accompanied by minimum fibre production. It is therefore postulated that potassium has a dominant effect in controlling both sugar and fibre production by flushing out tissue moisture and N. The multiple roles of K include sugar transport, protein synthesis and internal control of tissue moisture. Thus, K improves the fibre content of cane by gradually reducing tissue moisture (Fig. 10.5). Hartt (1929) provided evidence to show that K increases cutinization with decreased lignification. A lack of K results in an abnormal distribution of vessels with large vacuoles. Humbert (1975) suggested a heavy and late K application (even up to the 9-month stage) for maximum production of sugar and cellulose. He further recommended that a nutrient ratio of N:P:K of 2:1:2 be increased to 2:1:4 to achieve high sugar and fibre production.

Figure 10.5 suggests that late and heavy K application has a two-pronged attack and regulates both sugar and fibre. A sugar increase paralleling the fibre increase is evident following K addition. Firstly, K helps to convert reducing sugars to sucrose and improves the pol content of cane. Secondly, K reduces tissue moisture and aids conversion of thin-walled, large parenchyma cells to sclerenchymatous fibre cells. Exchangeable K, ranging from

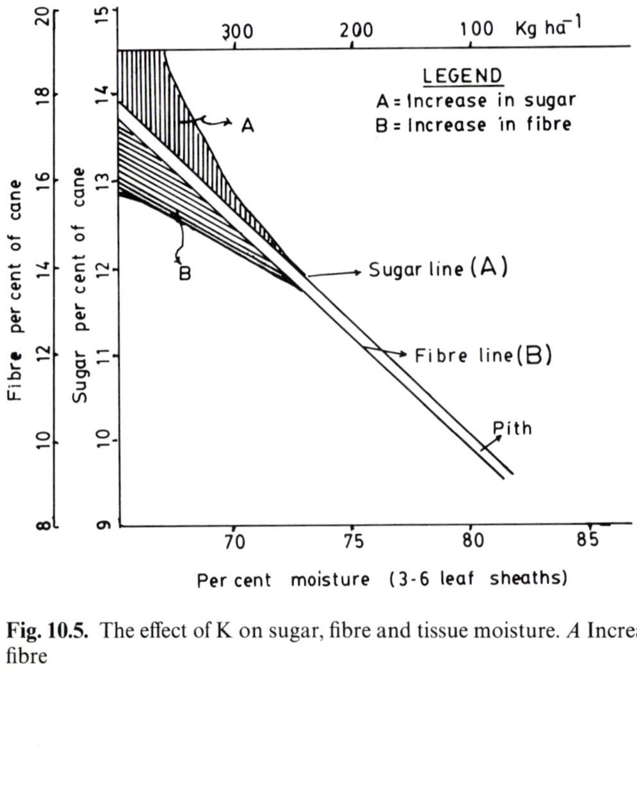

Fig. 10.5. The effect of K on sugar, fibre and tissue moisture. *A* Increase in sugar; *B* increase in fibre

65 to 125 kg K_2O ha^{-1}, is considered critical for sugarcane grown in most soil types (Hunsigi 1982a) and more than 250 kg K_2O ha^{-1} is desirable, coupled with late application, to achieve improvement of both sugar and fibre contents in cane.

The essentiality of Ca was established a long time ago since calcium pectate is a part of the cell wall and acts as a cementing material for cells. Calcium influences the conversion of parenchyma cells into fibre cells and, hence, Ca addition might improve the fibre content of cane. Van Dillewijn (1952) states that Ca is of particular interest since this element has the property of swelling cellulose and this process is associated with absorption of moisture. In other words, application of Ca reduces the fibre content with a concomitant increase in tissue moisture. However, field studies have not confirmed these trends. The long and short of it is that secondary and minor elements, like Ca, Mg, S, Si, Fe, Mn, Zn and B, have no perceptible influence on the fibre content or quality.

10.1.5 Proximate Analysis

Agricultural fibres consist of cellulose and hemicellulose (which together form holocellulose) along with lignin, pentosans and smaller quantities of fat, wax and mineral matter. Usually cellulose is incrusted with varying amounts of lignin, which is like a highly resilient plastic and protects the cellulose strands. Lignin is a heterogenous cross-linked polymer and is responsible for fibre hardness.

Three distinct types of fibres can be distinguished in sugarcane bagasse:

1. Rind fibres, which are long and made up of sclerenchymatous hypodermis and cortical tissue.
2. Fibrovascular bundles, which are relatively short and made of phloem and xylem vessels.
3. Pith parenchyma cells which are devoid of a fibrous structure.

Rind fibres constitute about 55%, fibrovascular bundles 20% and pith about 25%. Pith has a value as fuel or fertilizer but needs to be removed to upgrade the bagasse for pulping. The disadvantages of pith in bagasse during pulping are numerous and include the following: (1) it is non-fibrous, (2) it absorbs moisture, (3) it has a higher consumption of alkali during cooking, (4) it is slow in drainage, (5) it creates dirty specks on paper, and (6) it has a higher silica content. The proximate analysis of bagasse in comparison with bamboo and coniferous wood is given in Table 10.3 (Atchison 1982).

The cellulose contents were higher in coniferous and deciduous woods with negligible ash and silica. Hence, these woods are considered the best raw material for paper pulp (Table 10.3). However, the higher contents of pentosans in bagasse and bamboo improve the strength of fibres.

Table 10.3. Proximate analysis of bagasse, bamboo and various woods. (Atchison 1982)

Raw material	Cross and Bevan cellulose (%)	Cellulose (%)	Lignin (%)	Pentosans (%)	Ash (%)	Silica (%)
Bagasse	49–62	32–44	19–22	27–32	1.5–5.0	0.7–3.5
Bamboo	57–66	24–43	21–31	15–26	1.7–4.8	0.69–0.8
Coniferous wood	53–62	40–42	26–32	7–14	1.0	—
Deciduous wood	54–61	38–49	23–30	19–26	1.0	—

10.1.6 Varietal Differences in Fibre Content

The fibre in cane is only of structural importance. Among the different species of *Saccharum*, *S. robustum* and *S. spontaneum* possess a higher fibre content (Bull and Glasziou 1963) while *S. officinarum* has the lowest fibre content; interspecific hybrids and commercial varieties have intermediate values (loc. cit.). The allied genus of *Miscanthus* has the highest fibre content (51%).

Varietal differences in respect of fibre content were recognized in the early 1950s and varieties like CO421, CO413 and CO426 were rejected on account of a high fibre content since they caused ‘slippage’ in the mill. Most of the commercial varieties have 12–16% fibre and pol varies from 14 to 18%. CO421 has a relatively high fibre content with a fair amount of pol and still occupies large areas in central and eastern Africa. The commercial variety NCO376 has a higher fibre content (14.04%) than NCO310 (12.72%). An older cane of 22 months has a higher fibre content (15.35%) than a 12-month-old cane (13.38%). In general, flowered cane has more fibre than cane which is not flowering. Investigations with early rich canes like COC671, KHS3296 and CO997 indicated that fibre and sugar in cane are not incompatible and a variety can exhibit both increased fibre and sugar, which is definitely a desirable trait. Similarly, Q49 and B37172 are relatively rich in fibre and sugar. It has also been suggested that less than 13% fibre is not desirable since the mill does not get sufficient fuel for the boiler and additional fuel like oil or wood needs to be used to run sugar mills in Asia, Africa, etc. Infestation by pests like the top shoot borer is associated with the fibre and silica contents: the higher the content, the lower is the incidence of this pest.

Barnes (1974) has reported that the Hawaiian varieties differ in their fibre properties. The Hawaiian culture H37-1933 had an average fibre length of 1.82 mm compared with 1.22 mm in H32-8560. On similar lines, Hunsigi et al. (1986) found that B37172 has more pith or soluble matter while CO62175 has more useful fibres (Table 10.4).

There is intensive ongoing research into evolving varieties suitable for pulp and paper production at the Sugarcane Breeding Institute (SBI),

Table 10.4. Varietal differences in fibre content of bagasse. (Hunsigi et al. 1986)

Parameter	Variety			
	CO62175	CO7219	B37172	KHS3296
Useful fibre (%)	48.2	46.7	47.2	47.1
Pith and fibre (%)	25.5	26.5	28.5	21.8
Soluble matter and dust (%)	26.3	26.8	23.9	31.1

These values are relative since the cane was crushed in a small mill.

Coimbatore. The exploitation of Indo-American clones (IA) seems to be admirably suited for this purpose (Naidu 1986). The IA clones are the first or second nobilization products developed at the SBI from hybridization involving wild *S. spontaneum* and Puerto Rican or Hawaiian varieties (Anon. 1986). The fibre content in different groups of genotypes is given in Table 10.5.

It was further reported that from the first nobilization products of IA clones, nine types were selected and about 100 kg bagasse was sent for evaluation to Seshasayee Papers and Board. One of the clones, NS257, yielded more fibre and gave about 15% more pulp than normal bagasse obtained from commercial varieties. Trials are continuing at the SBI, Coimbatore, to test the progenies obtained from crosses involving IA clones and Coimbatore canes/*S. robustum* for pulp production. Naidu (1986) admitted that the present-day commercial varieties have undesirable traits like a powdery nature, larger amount of pith and small fibre cells. Hence, a separate programme was launched to utilize wild species like *S. robustum*. Naidu (1986) appears to have achieved success and some varieties developed at the SBI, Coimbatore, combine both high fibre and high sucrose making them well suited for use in sugar-paper-complex factories. Naidu (1989) emphasized that the genotype G354 (an interspecific hybrid of COC671×F1 of *S. officinarum* and *S. robustum*) showed considerable promise with high contents of both fibre and sucrose.

On milling certain varieties exhibited 'slippage' which is due to lignification in the rind portion. Hence, proper cane preparation and shredding are essential to achieve better juice extraction. An approximate measure of the

Table 10.5. Fibre content in different groups of sugarcane genotypes. (Anon. 1986)

Group	Fibre (%)	
	Mean	Range
S. officinarum × *S. spontaneum*	16.64	10.36–22.64
IA clones	20.57	18.62–23.01
Intervarietal hybrids	16.32	13.58–19.78
S. barberi and *S. sinense*	16.42	11.16–19.63
S. robustum	23.55	22.30–24.79

fibre and brittleness in sugarcane can be obtained by measuring the snapping stress (Hunsigi 1991).

10.1.7 Handling and Storage of Bagasse

Bagasse is a loose, bulky, porous material. The bulk density ranges from 120 to 180 $kg\,m^{-3}$. Extensive storage space is required and there is always a hazard from fire. Saved bagasse or the entire bagasse is transported by trucks in 12- to 16-t payload. For easy handling standard bale dimensions ($70 \times 30 \times 30$ cm) are used. If the paper mill is adjacent to a sugar mill, then bagasse can be transported on belt conveyors or by pneumatic conveyors.

In his keynote address Atchison (1986) stressed that the economic feasibility of using bagasse depends on proper storage and effective depithing. In the early 1930s the classical method for storing bagasse was to stack it in bales of 125 kg each with 50% moisture. The bales were stacked in pyramid-shaped piles with space in between each pile. The outer surface of the bagasse stacks was treated with a preservative like boric acid and finally the stacks were covered with an asphalt-coated metal sheet. This allowed the bagasse to dry at a controlled rate and prevented excessive build-up of acetic acid which accentuated the fermentation of sugar and caused a gradual degradation of cellulose. Much research has gone into this area and many bioengineering techniques are available to stack bagasse with minimal effects on fibre properties.

It has been found that fresh bagasse is not that suitable for pulping. Furthermore, the seasonal nature of sugarcane cultivation makes it mandatory to store bagasse for 3–6 months. The biodegradation of bagasse has been thoroughly studied (Rangamannar and Ramaswamy 1986). Bagasse has a residual sugar content (about 2.5%) and a high moisture content which accelerates the fermentation process. The darkening of bagasse occurs because of fungal growth. Furthermore, a combination of high temperatures and acidity results in hydrolysis of cellulose with a consequent reduction in pulp yield. An exposed surface area, heterogeneity of the fibres in bagasse and tropical conditions favour microbial growth and colonization. Rotten bagasse has a microbial population of $5 \times 10^8\,g^{-1}$ (Rangamannar and Ramaswamy 1986). The sequence of processes involved in bagasse biodegradation are given in Fig. 10.6. A reduction in the pentosan and hemicellulose contents and a loss of fibre properties has been shown following biodegradation of bagasse.

10.1.7.1 Conservation of Stored Bagasse

Early attempts to conserve stored bagasse involved retarding the activity of cellulolytic microorganisms by spraying bactericides like SO_2, formaldehyde, Na_2CO_3, etc. The biotechnique involves promoting mesophilic noncellulolytic bacteria which are acid-forming. Attempts should be made to

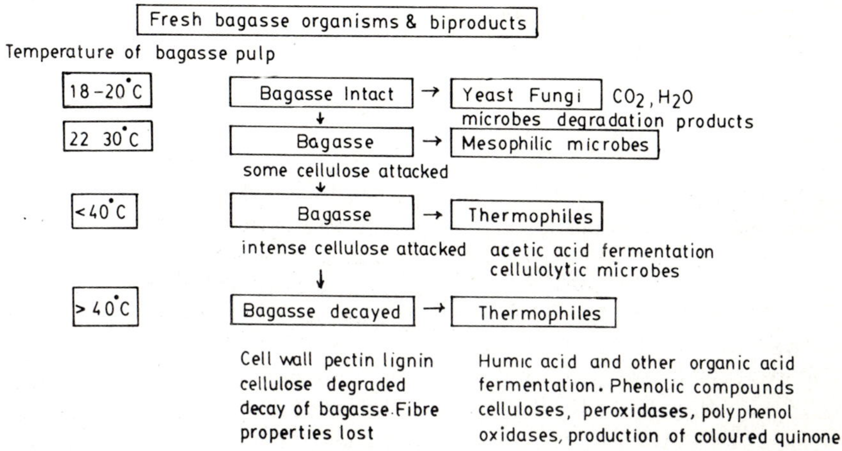

Fig. 10.6. Biodegradation of bagasse

create aerobic conditions and produce lactic acid with a pH of about 4. Thus, the general principle of conserving stored bagasse is to create an acidic, aerobic environment with proliferation of lactobacilli.

10.1.7.2 Development of Wet-Pile Technology

This is essentially the Ritter process, developed by E. A. Ritter, and involves the impregnation of moist or wet, depithed bagasse with a biological liquor. The seed lactobacilli culture is employed and the organisms are encouraged to multiply by the addition of nutrient media and molasses.

It was observed that fibre preservation was more effective in depithed bagasse than in whole bagasse since the former has less sugar. Thus, bagasse stored in a bioliquor has a higher pulp yield and lower consumption of chemicals (i.e. a lower kappa number) as well as a better colour and fibre properties.

The storage of bagasse, as advocated by Dr. Cusi, is a simple system where baled bagasse is stored in ventilated stacks with sufficient air space. The pith cells present in the whole bagasse protect the cellulose and the moisture content drops from 50 to 30–35% in about 50 days. However, there are certain limitations to this method under tropical conditions due to higher ambient temperatures.

10.1.7.3 Delignification

The lignin content in bagasse proves to be a disadvantage as it gives an undesirable colour to paper, besides increasing the consumption of chemicals during both cooking and bleaching. Therefore, a selective delignification was attempted at the Fungus Research Institute, Kerala, India. A particular strain of fungus selectively attacks pith and lignin while protecting cellulose and hemicellulose (Ramalingashetty 1985). A still higher quality pulp for speciality papers (currency notes) can be achieved by dissolving hemicelluloses (xylans) to leave only the cellulose. This is accomplished by an *Actinomycete* strain, *chainia*, which produces the enzyme xylamase and selectively dissolves hemicellulose leaving cellulose intact.

More recently, in Cuba, the white rot fungus (*Phanerochaete chrysosporium*, K3) was found to selectively delignify bagasse.

10.1.7.4 Removal of Pith (Depithing)

Thin-walled parenchyma cells remaining after the extraction of juice from cane constitute the pith. This is non-fibrous and has 93% of the calorific value of whole bagasse. It is also estimated that 2 t moist pith is equivalent to 1 t coal as fuel. Pith is highly absorbent, occludes dirt and reduces drainage and the strength of fibres. Above all, more chemicals are required while cooking and bleaching. The success of bagasse utilization for pulp and paper making depends on the effective removal of pith, which comprises about 25–30% of the dry bagasse. Pith can be conveniently used in boilers and special furnaces.

Three types of depithing are available: (1) dry depithing, (2) moist depithing and (3) wet depithing. Dry depithing causes severe dust pollution and wear and tear of the hammer mills; hence this method is hardly ever used. At present, two-stage depithing is recommended to achieve uniform fibres; this involves moist depithing at the sugar mill and wet depithing at the paper mill. An ideal form of bagasse storage is to leave the depithed bagasse as a slurry on a concrete floor of 40 000 m^3 for about 8 weeks. Further washing of the slurry on a hard rock or bitumin floor will remove an additional 5–6% of pith.

Rydholm (1965) has stressed the importance of depithing. Depithed bagasse has a higher pulp yield, brightness, tensile strength and burst and tear factor with a reduced Kappa number as compared with whole bagasse.

10.1.8 Fibre Properties of Bagasse

The fibres in bagasse are heterogenous in character and difficult to pulp mechanically. They are short and comparable to hard woods like *Acacia*, *Eucalyptus*, *Sesbania*, etc. The average fibre length (l) varies from 1.0 to 1.5 mm and the width (d) from 0.019 to 0.022 mm, with an l/d ratio of 74 to 85. A high pentosan and low lignin content give bagasse better strength properties (Table 10.6). A well-depithed bagasse, stored wet, has a high opacity. It suggests that the pulp yield is considerably less than for *Eucalyptus* but the strength properties are far superior to those of *Eucalyptus* as evidenced by higher values of the burst and tear factor and breaking length (Table 10.6).

Table 10.6. Proximate analysis, fibre properties and pulp yield of sugarcane bagasse compared with a *Eucalyptus* hybrid. (Hunsigi 1989)

Parameter		Depithed bagasse	*Eucalyptus* hybrid
Proximate analysis			
Bulk density (kg m^{-3})		67.4	225
Pith (%)		29.4	—
Lignin (%)		19.58	27.9
Pentosan (%)		21.24	14.8
Ash (%)		1.94	0.43
Fibre properties			
Length (mm)	Maximum	2.87	1.20
	Minimum	0.56	0.20
	Average	1.20	0.59
Diameter (μm)	Maximum	50	20
	Minimum	12	5
	Average	23	10
	Ratio l/d	52.2	59
Pulp yield (%)			
Unbleached		49.6	78
Bleached		44.5	74.7
Strength properties (Bleached pulp)			
Burst factor		39.1	10.9
Breaking length (m)		6550	2560
Tear factor		39.9	22.7
Double folds		360	Not available

Pulp can be defined as a solid system in which cellulose fibres, together with non-cellulosic components, are dispersed. The primary purpose of pulping is to loosen and liberate the fibres. In mechanical pulping the fibres are separated by abrasion and rupturing the cementing non-cellulosic materials. Chemical pulping consists of dissolving the middle lamella by acid, alkaline or neutral liquors. The pulp yield ranges from 30 to 60% depending on the raw material used. Dissolution of lignin, the main cementing agent, either mechanically, thermomechanically or biochemically, is a prerequisite for liberating fibres from non-cellulosic components.

From bagasse all grades of pulp, from highly bleached chemical pulps to high yielding thermomechanical pulps, can be manufactured. Furthermore, bagasse pulp mixed slightly with long-fibred pulp can be used for the manufacture of all grades of paper, such as toilet, tissue, corrugating medium, printing and writing papers and newsprint. In view of the low bulk density of bagasse and the more open nature of the cells, a high liquid to solid ratio is required and provision should be made for agitation. Since bagasse fibres are shorter and softer they require relatively low amounts of pulping chemicals (low kappa number, K) and a shorter cooking time. Continuous digesters (Kamyr type) and spherical-type digesters with a continuous cooking system are preferred to the batch-type digesters.

10.1.9 Bagasse Newsprint

More than 100 years of research have gone in to making bagasse newsprint since the first attempt in 1856 by Henry Low of Baltimore (USA). In the spectrum of paper grades, newsprint is a low-grade paper and the low-priced sheet should be able to withstand the requirements of high-speed printing presses making 40 000 to 60 000 impressions per hour. At an FAO conference held in Tokyo newsprint was defined as: 'The term Newsprint will be used without restriction as to fibre composition, thickness, ash content, degree of sizing or furnish. It will then apply for any kind of paper capable of being run through a modern printing press and of producing an acceptable sheet of printed newsprint at a reasonable cost'.

Bagasse newsprint (BNP) is now receiving world wide attention. There is no rigid definition for BNP but it should contain the maximum amount of bagasse pulp (60% or more in furnish) which is technically feasible and economically justified. According to FAO specifications, newsprint should possess the following characteristics:

Tear factor	46
Opacity (%)	86
Brightness (% MgO)	50
Grammage ($g\,m^{-2}$)	48.8

The historical development of BNP has been succinctly described by Atchison (1983, 1986). The global capacity for bagasse pulp was only 0.12 million tons in 1950 but by 1986 it had risen to nearly 2 million tons (Atchison 1986). Bagasse pulp is now used in practically all grades of paper, i.e. packing, wrapping, printing, writing, toilet, tissue, towelling, corrugating medium, linear board and bleached board. It is therefore treated as the future fibre of the tropical/subtropical world where cane is grown in abundance. Early attempts to produce a successful BNP failed because 100% chemical pulps were used including a high percentage of bagasse chemical pulp in the fibrous furnish.

Conventionally, newsprint is manufactured with 85–95% soft wood mechanical pulp and 5–15% soft wood chemical pulp. The long-fibred chemical pulp helps reinforce and increase the inter-fibre bonding strength. A growing body of evidence indicates that a maximum quantity of mechanical bagasse pulp is a prerequisite for making newsprint. Atchison (1986) has stressed three cardinal points for making newsprint from bagasse. These are:

1. Moist and wet depithing should be used to remove the maximum amount of pith.
2. A good storage method is required to prevent excessive loss in brightness and degradation of fibre properties during storage.
3. There should be a high proportion of mechanical pulp in the fibrous furnish.

It seems reasonable to postulate that earlier attempts to use chemical or semi-chemical bagasse pulp failed to produce an acceptable newsprint sheet.

According to Atchison (1986) the formulation for newsprint would approximate to:

1. 60% bleached bagasse mechanical pulp.
2. 30% semi-bleached, semi-chemical or chemical bagasse pulp or a mixture of the two grades.
3. 10% semi-bleached long-fibred pulp.

Orgill (1987) stressed that bagasse pulping for newsprint is an evolving process. At one extreme, the Philippines is perhaps the only country which utilizes hardwood (mainly *Albizia falcataria*) for newsprint, and the furnish is comprised of 55% thermomechanical pulp and 45% chemical pulp. Atchison (1986) stated that another breakthrough in the commercial production of BNP is the total dispensation of long-fibred chemical pulp in the furnish. Future BNP should consist of more than 85% bagasse pulp, which includes 50–60% bagasse mechanical pulp and 25–30% bleached bagasse chemical pulp, and the remainder will be *Eucalyptus* chemical pulp. There will be no additives in the furnish.

Yet another advance was attempted with the Hawzell process developed by the HSPA. This involved simulating mechanical wood pulp by grinding bagasse in disc-type refiners. The product was similar to mechanical pulp and imparted good opacity, ink absorbency and printing surface to the newsprint sheet. Reaffirming this, Villamil et al. (1986) suggested that work under the Cuba-9 project showed that chemi-mechanical bagasse pulp (CMBP) could be produced by chemical pre-treatment with caustic soda and the mechanical process is completed in disc refiners. According to these authors the pulp quality is adequate for newsprint and other grades of paper. This author reiterates that upto about 80% bagasse pulp can be blended with upto 20% long-fibred mesta or kenaf pulp (*Hibiscus sabdariffa* or *H. cannabinus*) to produce an acceptable grade of newsprint.

To conclude, sugarcane bagasse is the future fibre for tropics and subtropics for pulp and paper making. Cane cultivars need to be dove-tailed to maximise both sugar and fibre yields. Agro techniques like irrigation, manuring and plant density need to be adjusted to ensure a low tissue moisture but high sugar and fibre contents.

All grades of paper, including newsprint, can be manufactured from bagasse pulp through suitable blending with long-fibred pulp. Innovative pulping processes like the peroxide alkaline process or the non-sulphur chemi-mechanical pulping process deserve large-scale commercial exploitation since they ensure minimal air and water pollution.

10.2 Energy Cane

Fossil fuel supplies (crude oil, coal, natural gas) are finite and a series of energy crises have left no option but to turn to non-conventional energy sources like biomass, which is renewable and non-polluting. Biomass is essentially plant material, ranging in form from algae to wood. The energy content is relatively uniform and is in the order of 7000 Btu lb^{-1} (4000 cal g^{-1} or 16247 J g^{-1}) to 9000 Btu lb^{-1} (5000 cal g^{-1} or 20889 J g^{-1}). This represents a heating value of one-half to two-thirds of that of coal. The major advantages are that it contains negligible sulphur and little ash, while the chief disadvantages are that it has a low calorific value and high moisture content and is diffuse so that handling, storage and transport costs are high.

Hall (1979) estimated that proven fossil fuel reserves amount to about 900 billion tons of carbon, which is roughly equal to the standing biomass. Each year plant photosynthesis fixes about 200 billion tons of carbon with an energy content of 3×10^{21} J. The ideal candidate species for biomass production are corn, sugarcane, sorghum, sugarbeet and cassava, where the energy input–output ratio ranges from 6 to 9. Sugarcane is the most promising energy crop and Brazil made history of sorts by producing 12.5 billion litres of power alcohol in 1989–1990. These fuel alcohols give a 'lean burn' since oxygenated fuels reduce the emission of noxious gases. However, the idea of using fuel alcohol has been deflated, since prudence and economics dictate that it will not be viable until the crude oil price is around U.S. $60 a barrel. Nonetheless, this is not a zero-sum game and points out the importance of biomass as an energy source.

10.2.1 Efficiency of Phytomass Production in Energy Cane

It is a common truism that sugarcane is one of the most efficient plants for converting solar energy to stored energy (sucrose). Only the giant *sequoia* tree (*Sequoia gigantea*) surpasses sugarcane in phytomass production (Humbert 1975). On average, sugarcane produces 35–90 Mg ha^{-1} dry matter. This author has harvested 60 Mg ha^{-1} of dry biomass for cane grown in red, sandy loam soils. However, the energy cane produces a much higher amount of biomass than some grasses and trees like *Eucalyptus* (Table 10.7).

The extended growing season and higher rate of leaf area production linked to specialized carbon synthesis (C_4 of malic and aspartic acids) are

Table 10.7. Biomass yield of some grasses and trees

Species	Biomass (Mg ha^{-1})	Growth interval (months)	Reference
Eucalyptus hybrid	29.4	24	Rajan (1987)
Eucalyptus hybrid	66.5	36	
Eucalyptus robusta	38.3	24	
Sordan 70 A (sorghum × sudan grass hybrid)	44.5	12	Alexander (1985)
Napier grass	65.7	12	
Energy cane (high-density planting)	103.7	12	
Sugarcane	60.0	12	

responsible for the high phytomass production in sugarcane as compared with other C_3 plants. However, differences in dry matter production, including the fibre of *Saccharum* spp., are mainly due to differences in leaf area rather than to differences in photosynthetic rate (Alexander 1973).

A breakthrough in biomass production seems to be in the offing (Anon. 1990a) if large quantities of the chloroplast enzyme ribulose bisphosphate carboxylase oxygenase (Rubisco) is produced. This is a key enzyme in CO_2 fixation, accounts for 50% of the protein in the green leaf and is burned during photorespiration.

10.2.2 Food vs Fuel Farming

Globally, the potential land area available for cultivation is 3419 million ha (Buringh et al. 1975) and no additional land can be made available for biomass production. It was feared that energy farming would possibly compete with food crops for scarce resources like land, water and chemical fertilizers, but this theory has been discounted since biomass is 'man's friend' and a companion for all seasons. In fact, Tudge (1988) warns that biomass production is sine qua non, not only for the survival of man but also for the survival of 'fellow species'. To obviate the spectre of a serious competition between food and fuel crops, Lipinsky (1978) suggested the integration of fuel production with the food, material and residue system. Agronomic packages such as close spacing, coppicing, ratooning and multiple cropping are available to increase yields and reduce the unit biomass cost.

10.2.3 Energy Cane: A Management Concept

Alexander (1985) emphasizes that 'energy cane is neither a botanical identity nor a cultivar'. It is essentially a management concept where a remarkable package of anatomical, physiological and agronomic features of cane are tailored towards growth. In energy cane an attempt is made to divert the monolithic sugar commodity to a multi-product commodity. The faster growth rate of energy cane vis-à-vis sugarcane after the late juvenile stage (8 months and beyond) is maintained till harvest. It is theorized that the long internodes of cane are like an 'open bag of sugar' where fermentable solids are available for a growth-oriented process. There is ample evidence to show that higher sugar and biomass yields are genetically compatible. Energy cane is amenable to closer spacing, early canopy closure, expansive tillering, high rates of fertilization, especially of N, prolonged ratooning and high tonnage. High rates of inversion and a loss of 2–3 units in net rendement is more than compensated by the gain in biomass (Alexander 1985). Other distinct advantages of energy cane are the biological weed control through luxuriant green foliage, a greater leaf area duration and protection from erosion and soil compaction.

A special feature of energy cane is the propensity for tillering, stool spread, biologically active crown leaves and extended proliferation of surface and subsurface feeder roots.

Alexander (1985) has stressed that the 'weed' characteristics of energy cane are desirable and are seen in the wild species of *Saccharum*, i.e. *S. spontaneum*, *S. robustum* and *S. sinense* or in the hybrids with higher plasma content from wild species.

10.2.4 Agronomy of Energy Cane

Information on the agronomy of energy cane is sketchy, but for more details the reader is referred to Alexander (1985). The issue centres around a high plant density and N fertilization. Spacing studies undertaken by this author have shown that the row spacings can be 0.45 to 0.60 m with a millable cane population of over 30 m^{-2}. High density has to be achieved through an increased number of plants within a row. Hence, almost double the sett rate (45 000–60 000 three-eye-bud setts ha^{-1}) has been recommended for energy cane compared with conventional sugarcane cultivation. Further evidence of a higher biomass yield of cane (cv. CP65-357) grown in Florida has been presented by Gascho and Shih (1981). They obtained the highest yield of biomass, sucrose, total sugars and fibre at a closer spacing of 0.5 m and the values declined as inter-row spacing increased from 0.5 to 1.5 m (Table 10.8).

The authors stressed that the primary juice from stalks would, on fermentation, yield about 10 000 l ethanol ha^{-1}. A dry biomass yield of 52 Mg ha^{-1} represents a burning value of 9.2×10^5 MJ ha^{-1} yr^{-1} while nearly 25 Mg fibre ha^{-1} would yield additional energy of 4.4×10^5 MJ ha^{-1} yr^{-1}.

Among the nutrients, N is the most limiting factor for energy cane as the cane responds to higher rates of N addition. Application of N is two to three

Table 10.8. Effect of row spacing on sugarcane yield and composition at harvest. (Gascho and Shih 1981)

Parameter	Spacing (m)		
	0.5	1.0	1.5
Fresh stalk (Mg ha^{-1})	144	94	90
Sucrose (Mg ha^{-1})	21.0	13.4	13.1
Fresh trash (Mg ha^{-1})	28	18	16
Fresh biomass (Mg ha^{-1})	178	122	106
Dry biomass (Mg ha^{-1})	52	34	32
Total sugar (Mg ha^{-1})	26.4	16.2	15.2
Fibre (Mg ha^{-1})	23.6	15.2	14.5
Ethanol (l ha^{-1})[a]	9360	6110	5850

[a] Fresh stalk yield ×65
Data from Riveria Sand, Florida; mean of 3 seasons, i.e. 1977–1979.

times greater than the recommended dose to sugarcane grown in tropical Asia, Africa and Latin America. The normal recommended level of application of N to sugarcane is 200–250 kg ha^{-1} but for energy cane it is 400–450 kg ha^{-1}. Three equal splits are used to apply the nutrient at 2, 4 and 8 months after planting. This author is of the opinion that N addition at planting is not essential since it encourages weed growth, and the sett (two- to three-eye-bud seed pieces) has enough food material for germination and for the initiation of shoot and root growth. Band application of P at planting is beneficial since it promotes root growth and tillering. Nitrogen addition at 6–8 weeks is mandatory as the plant enters the first phase of tillering at this time. The second increment is at the grand growth phase (4 months), which is also a peak period for N requirement. The third addition is at the 8-month stage. If the N index (3–6 leaf blades) is less than optimum, an additional dose can be given at the 9- to 10-month stage; however, luxury consumption and N losses need to be avoided. The last increment can be given along with irrigation water (fertigation). Experience has shown that at 6–8 months K dressing along with irrigation water improves the girth of cane stalks and the number of internodes. Commensurate with high fertilization the water requirement also increases. Methods of irrigation include ridges and furrows, border and overhead irrigation. The width of the border varies from 10 to 15 m, depending on the size of the irrigation stream. The furrow length is 6–8 m, depending on the soil type. Despite the high water requirement of energy cane, the water use efficiency is more due to the root profile development and a high plant development density (Alexander 1985). Energy cane requires a fine seed bed for an extensive root profile development and stooling. Minimum tillage or conservation tillage for energy cane seems contentious.

10.2.5 Energy Cane Cultivars

These should be responsive to intensive agricultural practices. Yields of biomass, total sugar and high test molasses (HTM) in energy cane are much higher than in sugarcane. Biomass accumulation continues even after the late juvenile stage in energy cane and hence the biomass yields are much higher (Fig. 10.7). The duration of energy cane can be 12–16 months. An ideal energy cane cultivar is one that yields a high biomass and moderate sugar content. This may be achieved by crossing *S. officinarum* with wild species like *S. spontaneum*, *S. robustum* or *S. sinense* or with distant relatives like *S. barberi* or *S. edule*, and Narenga (Alexander 1973). Natural crossing is strongly suspected among the wild forms of *Saccharum*, *Sorghum*, *Miscanthus*, *Erianthus*, and *Sclerostachya* (Grassl 1946). Historically, Venkatraman (1937) first attempted to cross *Saccharum* with *Bambusa*. Thus, it is technically prudent to have intergeneric and interspecific crosses and a large *Saccharum* germplasm pool can be gainfully employed in developing energy and fibre canes. Promising varieties such as CO62175, NCO310 and B37172 are amenable to energy cane planting. Other promising canes are PR80-8-2,

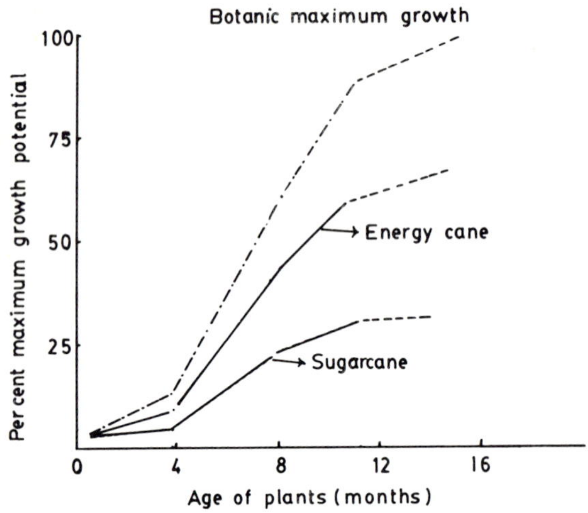

Fig. 10.7. The growth of sugarcane, energy cane and the hypothetical botanic maximum yield potential of *Saccharum* sp. (Alexander 1985)

B70-701 (F1 noble cane × *S. spontaneum*), PR980 (POJ2878 × CO281), PR64-1791 and CP65-357. According to Alexander (1990), cultivar US 67-22-2 of *S. spontaneum* is the most prolific producer of biomass. This is attributed to its expansive green leaf canopy, a high plant density (0.125–0.2 million ha^{-1}), rapid germination, minimum recumbency (lodging), proficient ratoon regrowth and extensive root proliferation.

At the SBI, Coimbatore, concerted efforts have been made to develop energy cane cultivars. Promising genotypes are (Table 10.9) EP18 and EP30 which are far superior in ethanol production at 12 months (Naidu 1989).

In a more recent study, Mishra et al. (1990) observed that a mid-late variety, CO7717 (4810 $l\,ha^{-1}$), was superior to early and late varieties like

Table 10.9. Some promising genotypes for alcohol production. (Naidu 1989)

Clone	Cane yield at 12 months ($Mg\,ha^{-1}$)	Pol % at		Fermentable sugars at		Calculated alcohol yield at 12 months ($Mg\,ha^{-1}$)
		10 months	12 months	10 months	12 months	
EP16	92.07	13.95	15.59	15.28	18.43	4.16
EP18	102.07	14.37	14.76	16.59	18.03	5.57
EP20	97.91	15.00	13.21	16.51	17.03	4.28
EP29	72.01	17.31	17.18	18.92	20.00	4.43
EP30	77.50	16.26	16.07	17.65	18.74	4.69
Standards						
CO419	62.50	14.80	12.86	17.80	16.02	2.63
CO62175	41.66	11.72	11.51	13.69	14.79	1.67
CO7704	66.66	18.48	16.16	20.01	19.61	3.27
COC671	64.16	20.36	15.61	21.46	18.07	2.84
COC8001	60.10	18.82	16.75	18.73	20.44	3.15

COJ64, CO7314 and CO1148 in alcohol production. Nitrogen levels and spacing regimes had no significant influence on either sugar or alcohol yield. However, a closer spacing of 60 cm and application of N at 180 kg plus 50 kg ha^{-1} as a foliar spray gave a numerically higher alcohol yield. The age at harvest had a profound influence on alcohol production and harvesting at 12 months (Mishra et al. 1990) gave the highest yield of 4810 l ha^{-1}; at 8 and 10 months the values were 3652 and 4782 l ha^{-1}, respectively. This author has noted that sweet sorghum can only yield about 2000 l alcohol ha^{-1} as against 4000–5000 l ha^{-1} produced by sugarcane.

10.2.6 An Ideotype of Energy Cane

Such an ideotype is conceived of as a luxuriant cane (Fig. 10.8) which is amenable to intensive farming practices including high rates of planting, N fertilization and watering. It is characterized by a propensity to tiller, early

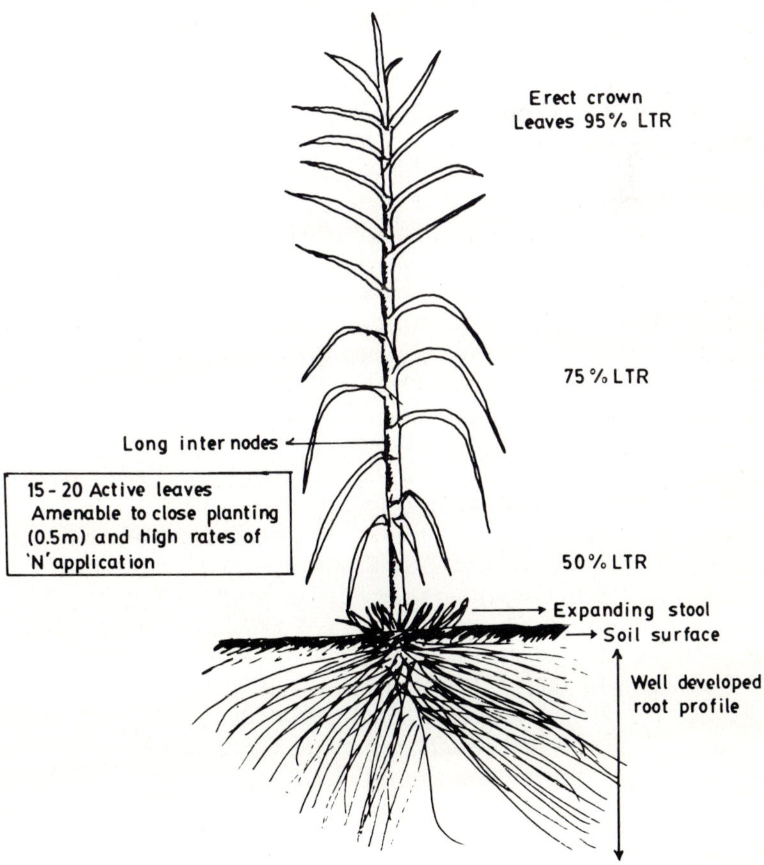

Fig. 10.8. An ideotype of energy cane. (Alexander 1985)

canopy closure, a high rate of stem elongation, ratooning and high tonnage with satisfactory juice quality. The active leaf numbers range from 15 to 20 (7-to 8-month stage) with the lower leaves being droopy while the crown has erect leaves. The total leaf surface per stool is much higher with a longer leaf area duration and the internodes are longer and number from 20 to 25 at harvest. An expansive root system (both surface and subsurface roots) is another prerequisite (Fig. 10.8).

10.2.7 Fuel Alcohols

Biofuels are the fuels of the future. Historically, ethanol (grain alcohol) has been primarily used as a beverage; it is produced from sugars (glucose, fructose) by the fermentative action of enzymes in yeast cells. Complex non-fermentable cellulose molecules are converted to glucose by the hydrolytic action of certain acids and enzymes.

Methanol (wood alcohol) can be produced from a synthesis gas which is a mixture of CO, CO_2 and H_2. Of the two fuel alcohols, methanol has a lower calorific value (5334 cal g^{-1}) than ethanol (7120 cal g^{-1}).

Methanol can be blended with gasoline in various proportions but there are attendant problems; upto 22–25% ethanol can also be blended with gasoline which is now popularly known as gashol.

10.2.8 Ethanol from Fermentable Solids

By far the most ambitious programme for the production of alcohol from sugarcane and other energy crops was launched in Brazil under the National Alcohol Programme (PNA or proAlcol). During 1985–1990, Brazil expected to produce 5–10 billion litres of alcohol with a total investment of about US $3.15 billion; upto 22–25% of this alcohol can be blended with petrol (Rosillo-Calle and Rothman 1986). A flow diagram showing the production of fuel alcohol from either primary juice or HTM or black strap molasses (BSM) is presented in Fig. 10.9.

Lipinsky (1978) even suggested a controlled production of crystalline sugar and if the sugar prices are low, substantial quantities of ethanol may be produced from the primary juice. It can be gleaned that HTM is distinctly superior to BSM and about 18 l HTM are equal to 22.5 l BSM. In fact, HTM is as good as the syrup or *meladura* (term used in Latin America).[1]

Yeast cells or their extracts are used to ferment molasses and alcohol is distilled off. Recent research suggests that the yeasts *Saccharomyces cerevisiae* or *S. uvarum* can be replaced by the bacterium *Zymomonas mobilis* for higher ethanol production from molasses (Rogers et al. 1983). The Biostil

[1] Syrup or *meladura* at the end of the last vacuum-boiling cells or bodies has 65% solids and 35% water.

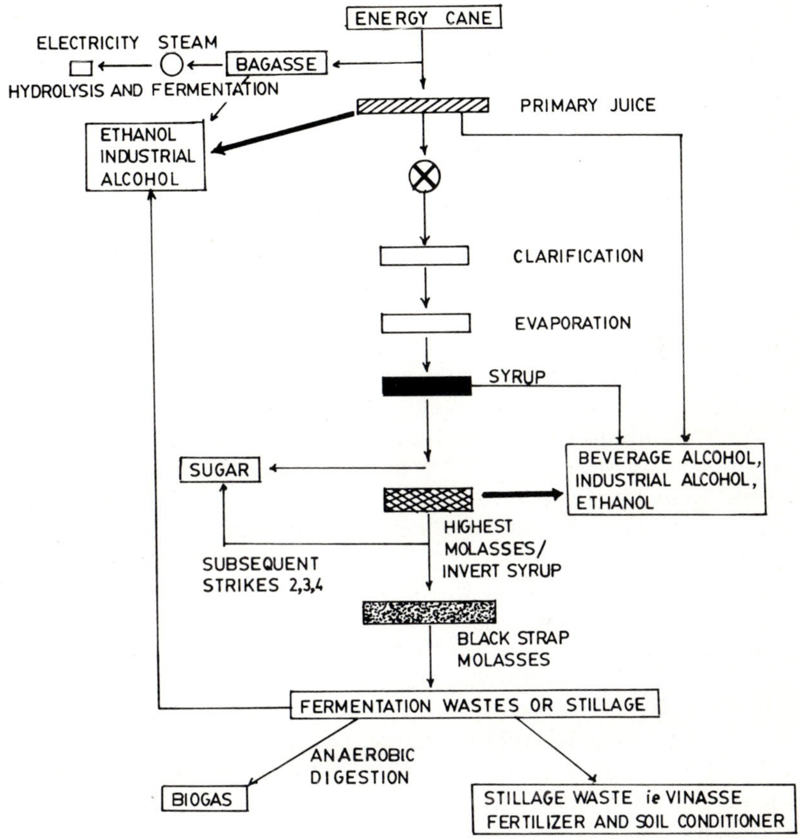

Fig. 10.9. Flow diagram to show the production of ethanol from energy cane

process seems to have the edge over other processes in ethanol production (Paturau 1986). This process was developed by Alfa-Laval of Sweden. Here, the fermentation environment is not limited by the final ethanol concentration but by the total osmotic pressure due to the accumulation of soluble non-fermentables in the fermenter. With this process only 60% of the conventional amount of processing water is required. Other advantages include minimal energy consumption and maintenance of high and stable yeast population (Paturau 1986).

The stillage or slops from the distillery poses environmental problems. The volume of stillage produced is about 13 times that of the alcohol produced and its BOD (biological oxygen demand) exceeds 25 000 ppm. Stillage disposal is achieved through fertilization, irrigation and biogas production. Vinasse, a distillery waste from fermentation is used as a soil amendment and fertilizer. Its application to Latosols has improved soil aggregation and increased the pH, organic carbon, base saturation and

CEC. It is a good source of K fertilizer and also supplies other minor elements like Zn, Fe, Mn and Cu.

10.2.9 Ethanol from Cellulosic Materials

Ethanol production from fermentable solids has overshadowed potential production from cellulosic biomass. Hydrolysis of cellulose can be accomplished with acids or the enzyme cellulase; the latter is cheaper and more cost effective. Lynd et al. (1991) have described the steps involved in ethanol production from cellulosic biomass through an enzymatic process. The energy output–input ratio of this system is 5 and the value is likely to increase when more efficient fermentative microbes have been identified.

In the fermentation process the quality of the biomass is equally important; hemicelluloses or lignins are not easily degraded. It is seen that sugarcane bagasse has a higher cellulose content than corn and Pruett (1981) has shown that mutant strains of *Clostridium* sp. are efficient converters of cellulose or hemicellulose to ethanol. More recently, Landisch and Svarezkopf (1991) stressed that the ability to ferment pentoses (xylose) is a key challenge and microbes like *Candida tropicalis* and *Pichia stipitis* are efficient in this area.

Lynd et al. (1991) have cautioned that the high costs involved in producing ethanol from cellulose are a major impediment to utilizing this technique. Nevertheless, production of ethanol from cellulosic biomass is a developing technology, although significant improvements are possible in the areas of pre-treatment, enzyme activity, production, recycling and developing energy-rich crops.

Chapter 11 Cane Development and Transfer of Technology

There is an old cliché, 'cultivate the grower and he will grow quality cane for the factory'. This implies an intimate relationship between the grower and the miller. Two scenarios exist: one where the factory owns the entire plantation and the other where farmers grow cane on owned or leased land and supply it to the factory. Thus, cane development requires specialized public relations involving both art and science. Management may be defined as managing people and, hence, the management of a cane department must essentially cultivate the growers to accept the philosophy of sustained growth of the factory and the welfare of a large sector of growers. The cane department has the onerous responsibility of motivating the cane growers to grow a particular variety or to adopt a particular agronomic package so as to maximize cane and sugar outputs.

11.1 Technology Transfer

When considering cane farmers, Blackburn (1984) emphasized the need for an intimate relationship between the farmer and the manufacturer. He observed that in Trinidad, Jamaica and Barbados, more than one-third of the cane is grown by farmers, while the rest is grown on estates owned by the factories. In Guyana, nearly 7% of the total cane comes from the farmer. In Sudan there are tenant farmers who rent the farm and supply the cane to the mills. In Australia and Fiji, estates are divided into smallholdings and these are entrusted to independant farmers to grow cane with technical assistance from the factory. In Barbados, a distinction is made between smallholdings and estates with a demarcation line of 4 ha. In India, Pakistan, Bangladesh and Sri Lanka most of the cane is grown by small, marginal and big farmers, though only a thin line divides these groups.

Technology transfer aims to increase the sugarcane area (without being detrimental to food crop production), yield per hectare and amount of sugar in cane. Since the sugar factory is the ultimate beneficiary it should provide critical inputs like seed, fertilizers and plant protection chemicals through an easy credit system. One FAO study has shown that the diffusion of innovations does not follow a definite pattern in developing countries. Hence, it has been suggested that there should be a direct channel between the farmer and researcher.

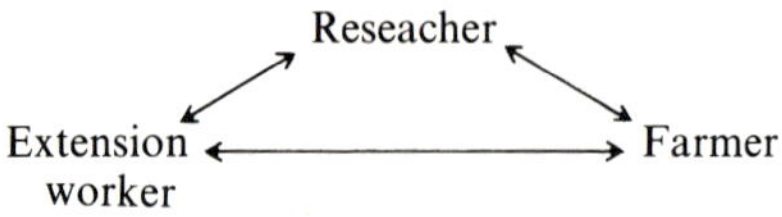

This would assure proper feedback and the extension agency could continue to liase between the farmer and researcher. However, it is only in the mainland China where the learning from farmers has been institutionalized through a farmer's forum and associates. The FAO defines 'extension education' as an informal out-of-school educational service to train and influence farmers and their families in the adoption of improved practices in crop and livestock production, management, conservation and marketing. There must be an overall change in the outlook of the farmer so that he is receptive to technological changes. In India great strides have been made through location-specific research under the National Agricultural Research Project and the Training and Visit system.

Technology transfer in sugarcane is relatively easy due to the special relationship between the farmer and factory. Ipsofacto, every effort should be made to achieve sustainable cane and sugar production with minimal

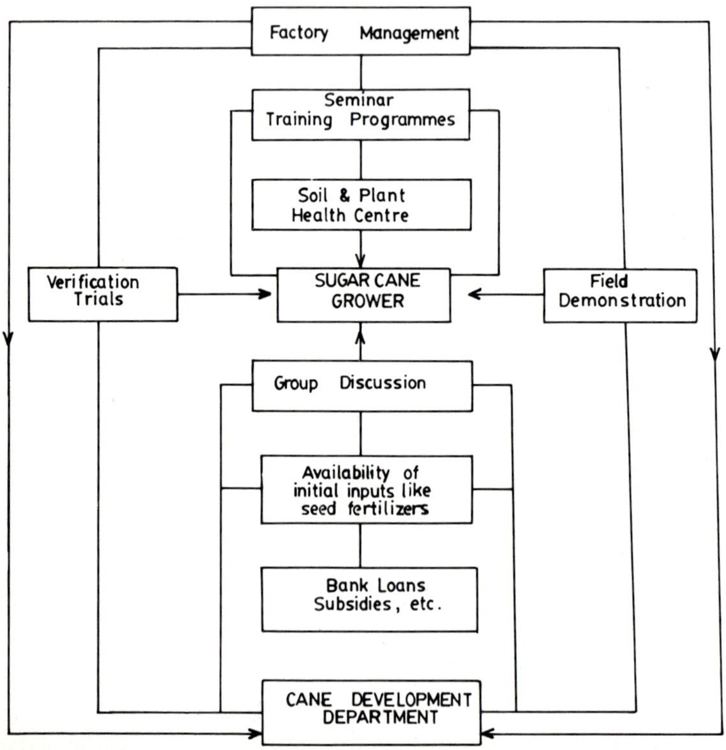

Fig. 11.1. To show how management reaches the cane grower

environmental pollution. Figure 11.1 shows how the factory can reach the cane grower with the ultimate objective of augmenting sugar and cane yields. Jagannath Rao and Narasimha Rao (1990) emphasized that the primacy should be given to group discussion and field demonstrations.

11.2 Methods of Cane Purchase

It is beyond the scope of this book to describe the different methods of cane purchase used throughout the world. However, some broad guidelines are given. There is no universal method to indicate cane purchase. Nevertheless, each system should take the cane quality into account. Stale cane is avoided and freshly cut cane arrives at the mill within 18–24 h. Burnt cane or frozen cane is usually rejected or the supply of such cane is heavily penalized. Manually cut canes are cleaner with less trash, roots, adhering soil, etc. The tops are removed and allowance is made for up to 2% trash and extraneous material. Meade and Chen (1977) point out that cane purchase discussions between the grower and miller revolve around three parameters: the general system of payment, evaluation of cane quality and the proportion of payment for by-products. The simplest system is the flat price, based on the price of raw sugar. This system takes into account the price of molasses and may include other by-products. Guyana and Trinidad use the flat rate system. In India, a base price is fixed at 8.5% of sugar recovery and every increase in unit sugar bagged will get a proportionally higher cane price. Some important methods of evaluating cane include the following.

1. The Java ratio (JR) is the arbitrary milling ratio, i.e. sucrose in cane or sucrose in crushed juice × 100. JR does not take into account the different fibre content of different varieties (Meade and Chen 1977).
2. The SJM formula proposed by Deerr postulates that for a given juice of J purity and producing sugar of S purity with molasses of M purity, the percentage of total sugar (pol) in the original material to go into sugar will be:

   $$SJM = 100\ S(J - M)/J\ (S - M).$$

   This formula envisages that the pol in the original material separates into that which goes into sugar with the remainder going to molasses. It does not take into account other losses.
3. The pol ratio (PR) is the direct measure of cane quality expressed as tons cane (TC) /tons sugar (TS).
4. Arceneaux's universal equation is given as:

   $$S' = Sx - By,$$

   where S' is the available sugar (%), S is the pol (sucrose) in juice, B represents the brix of primary juice and x and y are the factors connected with fibre content of cane as well as brix and pol of juice.

5. Australia was perhaps the first country to make payment for sugar rather than cane and the grower is rewarded according to the sugar content of the sugarcane. The Australian commercial cane sugar (CCS) formula is given by Meade and Chen (1977) as:

$$CCS = 3P/2\ (1 - [F + 5]\ /100) - B/2(1 - [F + 3]\ /100),$$

where P is the pol percentage of the first expressed juice, B is the brix percentage of the first expressed juice and F is the fibre percentage in the cane.
6. The Winter and Carp formula is widely accepted throughout the sugar world but has undergone many modifications (Meade and Chen 1977). The formula is:

$$x = S(1.4 - 40/P),$$

where x is the available pol (sucrose) percentage in cane, S is the percentage pol in juice and P denotes the purity of juice.
7. The Louisiana method used by the US Department of Agriculture envisages a 'standard cane' having 12% pol with a juice purity of 75%. Thus, a premium is added or penalty deducted when the 'normal juice' (first expressed juice or laboratory mill juice) exceeds or fails to reach this quality (Blackburn 1984).
8. In Spain a simplified yield of commercial sugar (Y) is computed by assuming an average of 0.779 (Barnes 1974):

$$Y = 0.779\ P - 3.05,$$

where P is the pol percentage of the first expressed juice. The loss during processing is assumed to be 3.05%.

Once the quality of the cane has been assessed, payment to the grower and miller is made in the proportion 60:40; this method is used in the Philippines. In other countries like Mauritius, Réunion and Jamaica 75% of the sugar value is paid to the grower along with receipts in scums and molasses. In Thailand and Indonesia a tripartate agreement is reached between the miller, grower and government with the latter acting as an umpire. One technologist humorously remarked that in a boxing bout the umpire receives more punches than the boxers.

11.3 Seed Cane

True seed from fluff is used only in breeding stations for producing new sugarcane seedlings. Commercial propagation is through setts or cuttings of stems. Setts with two or three eye buds are used to break apical dominance and assist germination. A good seed is one which is free from pests and diseases, high in moisture content and rich in invert sugars and soluble nitrogenous compounds. Hence, young cane of 8–9 months, well manured and irrigated is ideal seed material. The practice of using the top portion of cane (unarrowed) for seed purposes is a contingency plan. However, a mature cane (12–14 months) can be used under drought or saline/alkaline conditions. This author reiterates that ratoon cane should never be used for seed.

Information on the agronomy of seed cane is very sketchy. However, an additional 35–50 kg N ha^{-1} and 25–30 kg P_2O_2 ha^{-1} is recommended for the seed crop. This can be applied as late as 6–8 months after planting.

Humbert (1968) suggests prefertilizing seed cane, i.e. fertilizing 4–6 weeks prior to harvest. The prefertilized setts germinate rapidly with vigorous seedlings and a high proportion of roots and shoots.

The term 'short crop' (nursery crop) is used in Java (Indonesia) and such a crop is used to resist the spread of the serious 'sereh' disease. A carefully tended nursery is maintained with adequate watering and manuring and the crop is harvested after 6 months for planting. This is followed by another cut 6 months later. Such a practice helps to increase the rate of seed propagation. In seed cane the rate of propagation is low, i.e. 1:10 or 1:12 (a multiplication rate of 10 to 12 means that 1 acre of seed cane will be sufficient for 10–12 acres). But in a short crop, the multiplication rate can vary from 25 to 100 times.

According to Van Dillewijn (1952), the *seblang* or sprouting method gives the highest rate of sett propagation. Under adverse edaphic and climatic conditions, sprouted setts or slip setts would assure a comfortable plant stand. In this method the seed crop is grown in light soils which are well fertilized to promote profuse tillering. As the tiller with a root system develops, it is severed from the mother shoot and planted. The succession of tillers formed are, in turn, themselves used for planting. The success of this method depends largely on the regenerative power of the plant, which depends primarily on the variety. The *seblang* method is ideal for gap filling in ratoons or in plantations where the plant stand is poor.

Another variation to the *seblang* method is the *rayungan* or *rajoeing* method, in which 6 to 8-month-old stalks are decapitated while in the field to break apical dominance and side shoots develop in about 4 weaks. Each shoot with its attached internode is severed from the mother cane and planted. Thus, rapid multiplication of up to about 100–200 times is ensured. Ample irrigation and fertilization are necessary to ensure the full development of *rayungans* on the stalk.

There are a number of sett-borne diseases, such as RSD, GSD, smut, mosaic, red rot and leaf scald, which adversely affect the seed cane quality. Heat therapy is considered to be the panacea for most sett-borne diseases and heat treatment not only controls diseases like GSD, RSD, smut and to some extent red rot but also arrests varietal yield decline and rejuvenates the commercial varieties. Thus, growers should establish nurseries with heat-treated cane but the ideal condition is for the factories to provide heat-treated seed material. Heat treatment can be done with either hot water(52 $\pm$ 1°C for 2 h), hot air (54 $\pm$ 1°C for 8 h) or moist hot air (54 $\pm$ 1°C for 4 h). A moist hot air or hot water treatment seems to be the best as there are few

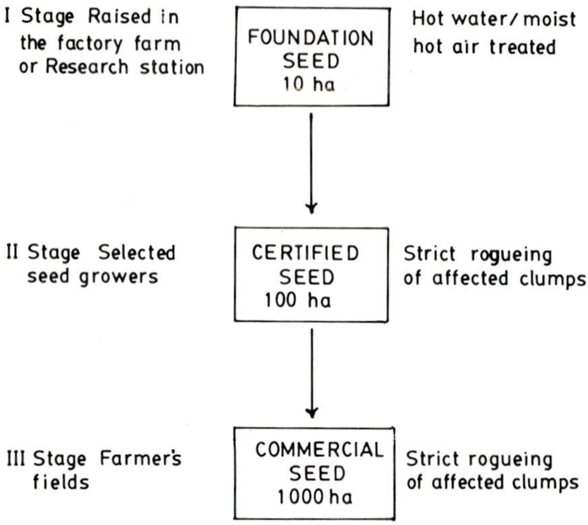

Fig. 11.2. Three-tier seed production programme

escapes.[1] With a hot air treatment there is more drying of the cane and a considerable loss in germination. In any case, strict roguing is an essential prerequisite to obtain healthy seed material.

A systematic three-tier seed programme is suggested (Fig. 11.2). The first tier is the foundation seed where the crop is raised from heat- treated seed pieces. The second tier is the first multiplication, called certified seed, and this is raised from the heat-treated primary foundation seed stock. The third tier consists of the commercial seed raised from certified seed and this commercial seed is given to the farmer for general cultivation (Fig. 11.2). Heat treatment is used only for raising foundation seed and not for the subsequent two multiplication stages. Be it as it may, strict supervision of the production of seed free from secondary infection is essential for the success of the three-tier seed programme.

[1] Despite heat therapy, some setts show disease symptoms upon germination, hence, the term escapes.

Chapter 12 Sugar and By-Products from Sugarcane: Recent Developments in Manufacturing Techniques

By Mydur Anand[1]

As this book focuses mainly on cane, only the salient features of the manufacturing operations associated with cane sugar will be explained, with emphasis being placed on recent developments (see Fig. 12.1).

12.1 Essential Steps in Manufacture

In this chapter explanation of the manufacturing process will be confined to the aspects and activities involved after the receipt of harvested cane at the sugar mill and until the product 'sugar' is produced and stored, i.e. from

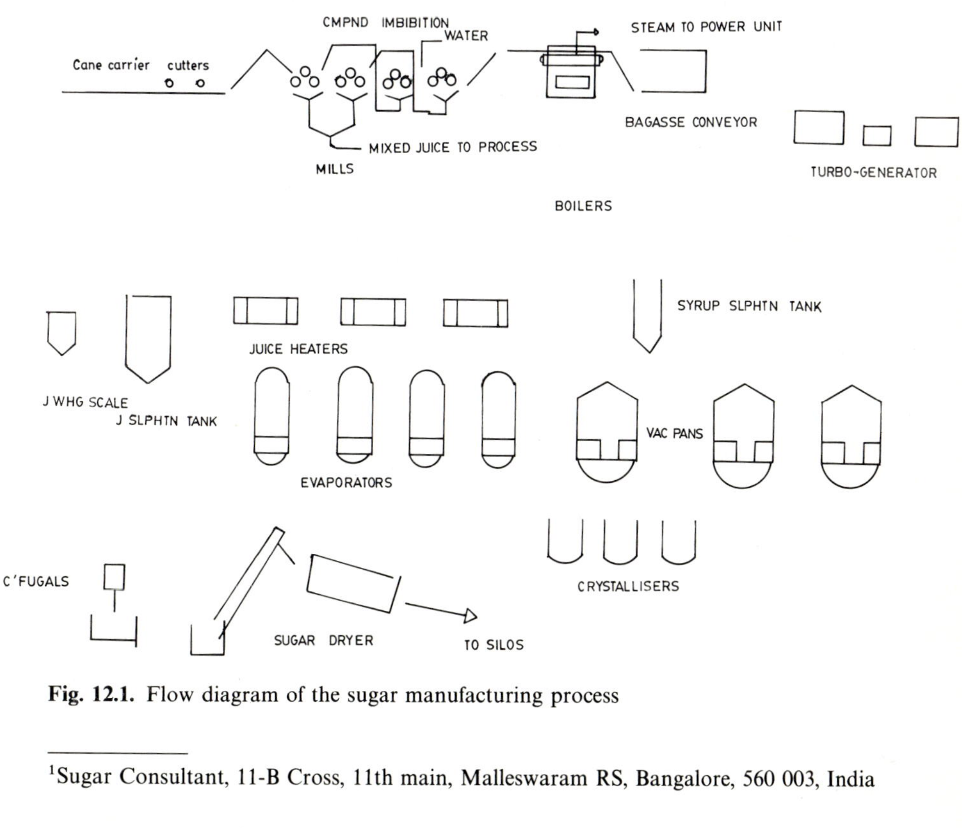

Fig. 12.1. Flow diagram of the sugar manufacturing process

[1]Sugar Consultant, 11-B Cross, 11th main, Malleswaram RS, Bangalore, 560 003, India

handling of the cane to handling and despatch of the sugar. Mention is also made of the varieties of sugar produced in various grades and packages.

The essential steps involved are:

1. Handling of cane, including weighing and sampling.
2. Cane preparation and milling.
3. Clarification of the juice.
4. Evaporation and concentration.
5. Syrup/melt treatment.
6. Sugar crystallization and massecuite handling.
7. Storage, packing, etc.
8. Steam and power production including bagasse handling.

12.1.1 Cane Handling

Sugarcane is transported to the sugar factories from cane farms (owned by the factories themselves or by individual farmers) by various transport systems. The cane has to be unloaded, stored for a period, and fed into the cane-carrier which conveys the cane through a series of 'cane preparation' devices (cutting cane into shreds to enable easier extraction of juice in the mills) before it enters the mills.

12.1.1.1 Delivery of Cane to the Mill

Different countries use different systems. The two operations, transport and handling of the harvest, are to a large extent interlinked, but in practice multiple methods of transport may be associated with any one method of harvesting. The harvesting, loading and transport systems in vogue may be grouped as shown in Table 12.1.

Table 12.1. Systems involved in harvesting and transporting sugarcane

Harvest	Loading cane into transport vehicles	Transport systems
Manual		
1. Green cane	Manual	Bullock-drawn carts
2. Burnt cane	Mechanical	Tractor-drawn trailers
(whole sticks)		Trucks
		Rail trolleys
Mechanical		
1. Whole sticks	Mechanical	Tractor-drawn trailers
2. Chopped billets		Trucks
		Rail trolleys

12.1.1.2 Cane Weighment

Each transport vehicle is weighed individually to ascertain the net weight of cane brought into the mills.

At present, electronically operated weighbridges are in vogue, even in developing countries; these enable the weighing of data (weight, interface/suppliers' name, weight of cane, etc.) to be documented both for the mill accounting system and for the cane suppliers' accounts, regardless of whether the supplier is a nuclear estate or an outgrower. In this way an assessment can be made of quality, price, amount payable, etc.

12.1.1.3 Unloading Systems

These are largely linked to the transport systems.
For whole stick canes the unloading systems are:

1. An overhead gantry with travelling hoist to offload the cane using a sling system or a mechanical cane grab.
2. Tilting of either road or rail trucks and trailers using hydraulically operated platforms; either side unloading or rear unloading is possible.
3. Chain net unloaders. This involves a complete system, i.e. trucks and trailers must be designed to have the chain net system. Chopper-harvested cane may also be unloaded using this system.

Chopper-harvested cane is usually transported in bin-type containers, either by road or rail. These are unloaded into a cane carrier at the mills, either by tipper or dumper trucks.

12.1.2 Cane Sampling

Although in many countries a basic cane price is fixed by the authorities, the actual price varies depending on the quality (recoverable sugar) of cane supplied. This has led to the development of sophisticated equipment like core samplers, mounted on booms or specially designed tractors, for continuous sampling of the primary juice.

Core sample canes are subjected to mixing, hydraulic juice extraction and automatic pol determination steps.

In some countries, cane is paid for simply on a weight basis. The basic price is determined by the average sugar recovery obtained by the mill during the crushing process. All the cane suppliers are paid the same rate. The weakness of this system is that payment does not take into account the quality of cane supplied by any individual farmer, but it is a workable system where the number of suppliers is too large to enable an individual sampling system to be used. There is also some equity in this method because the factory pays, in total, for the sugar content of the cane crushed in a particular season, and the farmers share the price equitably.

It also needs to be emphasized that even with sophisticated methods of sampling and analysis, disparity is still observed between the estimated sugar recovery, as indicated by sampling, and the actual sugar recovered during a particular period.

12.1.3 Cane Preparation

12.1.3.1 Cane Laundering

In some countries the cane is harvested after the standing cane crop has been burned to eliminate cane trash, dried leaves, weeds, etc.; this enables a worker to harvest more than twice the amount of cane he would otherwise harvest, i.e. green cane. However burnt cane deteriorates faster and needs to be crushed soon after harvest.

With chopper-harvested cane, a green-cane harvest has been made feasible and is very popular.

Where cane burning is used, such cane carries a lot of soot and other extraneous matter. Also, in areas where rains are frequent, even during the crushing season, a lot of wet clay sand can stick to the cane. To eliminate as much soot and extraneous matter as possible, cane laundering is carried out; this involves washing the cane with copious quantities of water. Obviously, this consumes enormous quantities of water and energy, the latter being required for pumping, recirculation, effluent treatment, etc.

As with any system, there are some inherent disadvantages. The water used for laundering gets polluted, and the microorganisms involved may stick to the cane so that the infection is carried through to the juice extracted at the mills; this may result in a considerable loss of sugar.

12.1.3.2 Cane Disintegration

Cane is a hardy crop; the cells containing the sugar juice are held between protective, wiry fibres. Juice (sugar) extraction is feasible only when these juice cells have been exposed and ruptured.

A variety of gadgets are used for such disintegration, and quite often two or three devices are used in series. Prominent among them are rotary cane-cutting knives and high-speed disintegrators with rigid or swinging arms (unigrators, shredders, mincers, etc.); these may have heavy and hard hammer-type heads which beat the cane against steel bars that serve as anvils.

The greater the extent of cell rupture (which is measured and serves as an index of cane preparation), the higher is the sugar extraction.

Until recently, attempts were aimed at high cane preparation but without reducing the cane to dust (i.e. a 'woolly' preparation was not desirable); long fibres were retained intact to aid proper mill feeding.

However, lately, with the introduction of pressure feeding (PF) devices, such as Donnally chutes and forced feed rollers, resulting in five-roller and six-roller mills, even a 'woolly' preparation is acceptable, as in this way a large number of juice cells are definitely opened.

In recently installed milling systems, the power consumption in the cane preparation stage is about 40–50% of the total power consumed in the milling operation; a few decades ago it used to be a very small fraction.

12.1.4 Milling and Juice Extraction

The mill is the most important station in the entire sugar cane factory, in all aspects, such as total investment on equipment, energy consumed and role in determining the overall efficiency, skill and costs involved in the operation and maintenance, etc.

Diffusion is another parallel system of juice extraction. It is the process universally adopted for sugar beet, but its application to cane has not been very successful. This system claims to use only about 60–75% of the energy used in conventional milling and the efficiency achieved is as good as, if not better than, that achieved in milling. However, this system requires excessive quantities of water, resulting in additional requirements for steam and fuel. Its main weakness is its lack of flexibility.

While mills, originally designed to crush, say, 100 t cane h^{-1}, have been upgraded to crush almost 200 t h^{-1}, the diffusion systems cannot be upgraded, except perhaps by about 10–20%, in spite of improvements in cane preparation, PF systems, grooving patterns, roller arcing, etc. Hence, in spite of the high costs and energy consumption, milling continues to rule supreme and diffusion is no longer an option, even in new installations. Therefore, further comments on juice extraction will be limited to milling only.

12.1.4.1 Mill Construction

Basically, the three-roller configuration, adopted nearly 200 years ago, holds ground today. But the improvements and modifications to both the system as a whole and in detail have been phenomenal. They have resulted from improvements in general engineering such as the development of steam turbines and electrical motors, developments in metallurgy resulting in the use of cast steel, precision-engineered power transmission, improvements in welding techniques so generally used in roller arcing and rebuilding of worn-out parts, improved lubrication techniques and better hydraulic systems, etc.

12.1.4.2 Three Roller-Six Roller Mills and Pressure Feeding

The conventional three-roller mills suffered from 'slippage' of the cane at the feed site, in spite of 'chevron'-type grooves. A fourth roller was added to improve the feeding characteristics and a further improvement was made by providing a tall, closed chute, usually called a Donnally chute, about 3 m high; this supplied cane at a higher bulk density, making it easier for the mill to accept the 'feed'. In this way both the capacity and efficiency of the mills was enhanced.

Another development of even greater impact has been the introduction of PF devices. These consist of a Donnally chute and pair of rollers to accept the feed and force it into an enclosed chute on the main mill. The feed is supplied at a much higher pressure, resulting in a doubling or more of the capacity. As the cane is well compressed in the chute, some juice is often also extracted. The juice wets and soaks every piece of cane, which improves the imbibition efficiency enormously, and thus helps improve the mill extraction efficiency in addition to enhancing the crushing capacity of the mill.

The two PF rollers effectively convert a three-roller mill into a five-roller mill. An under-feed roller, below the chute, is found to be helpful and this then makes the mill a six-roller mill; however, the six-roller mill has only two effective compressions.

The PF rollers used are mainly of two types: those with conventional grooved rollers (GRPF) and those with toothed segments (TRPF). There is not much to choose between the two; it is possible that TRPF was developed to compete with GRPF without infringing upon the patent laws.

It must, however, be admitted that the PF systems have resulted in the success of the milling system over the 'diffusion' systems.

The PF system enables a lower number of mills to be used for a given capacity; a five-mill unit with a PF system can crush more cane more successfully than a seven-mill unit of the conventional type.

Lately, some designs have been produced which, while being six-roller mills, provide four compressions per mill. Indeed, milling technology is progressing by leaps and bounds.

12.1.4.3 Low Pressure Mills

The conventional three-roller mill is a crude device for extracting juice and it is made cruder and uglier by its conversion to a five-roller or six-roller mill.

As a contrast to the PF system, this author proposes a simplified low pressure (LP) system, which essentially consists of intensive cane disintegration carried out in two stages, i.e. once before the two-roller first milling and once after the primary crushing. This is followed by a series of six two-roller mills and finishes with a conventional three-roller mill to serve as a de-moisturizing mill. The series of two-roller mills provides almost twice the number of imbibition stages used in the conventional three-roller mill with an equal number of rollers in the entire milling train. Thus, a 17-roller milling tandem in the LP system provides ten compressions and eight imbibition stages, while a conventional 18-roller mill tandem (three-roller type) would provide 12 compressions but only five imbibition stages.

The importance of imbibition has not been fully realized, but this imbalance is corrected in the LP system.

The LP system, originally proposed in 1982 by this author, was given a practical shape by Dr. W. Leipzig in 1989 when he converted some old mills in Brazil and Mexico to the two-roller mill system.

The advantages of a two-roller mill compared with a three-roller mill are:

1. Absence of a trash plate, which absorbs about 20% of the drive-energy.
2. No choke of mills.
3. The two rollers can easily be made to run at different speeds, improving the shredding effect.
4. Constant-speed drives, like AC motors, can be used.
5. There is an overall reduction in energy consumption, maintenance costs, etc.

12.1.5 Mill Drives

Mills are perhaps the largest energy guzzlers in sugar production. Fortunately, the cane is endowed with a large content of fibre which can be used as a source of energy to provide all the steam and motive power required to operate the sugar plant. Thanks to developments in technology, a lot of fibre (bagasse) remains as surplus after the steam and power needs have been met. This surplus bagasse is being gainfully employed in producing surplus power (co-generation) or paper or particle board, etc. With this scenario, the economical use of power to drive the mills has attracted special attention.

12.1.5.1 Types of Drive

The mills are driven at about 3–6 rpm, depending upon the size and capacity of the mill and the aptitude of the engineers. Regions like Louisiana (United States) adopt higher speeds to achieve higher grinding rates, but in Australia and other countries the mills are run at lower speeds to improve juice drainage and hence juice extraction.

Steam engines running at 40–100 rpm were in vogue early in the century but these gave way to high-output, high-speed steam turbines by about the middle of the century. With a higher pressure of steam, the turbines were designed to have much higher horsepowers than was possible with steam engines. Turbines are trouble-free and easy to maintain, unlike the reciprocating engines. Turbines are also compact units, but as they run at high speeds of 6000 to 10 000 rpm a number of speed-reduction steps are needed. Helical gears are used at higher ranges of 10 000 to 1000 to 50 rpm and spur gears are used at lower speed ranges, i.e. from about 50–5, at the roller drive end. Turbines have many desirable qualities: simplicity of operation, higher energy outputs at higher thermal efficiencies, ease of speed control by use of remote and automatic instrument systems, capacity to absorb impact loads and they can be used with variable loads, high starting torques, etc.

Today, steam turbines are the most popular devices to drive mills.

12.1.5.2 New Drive Systems

Lately, hydraulic drives and DC motors have made an appearance as these are better devices for driving mill.

Hydraulic Motors. These primarily consist of a power pack where hydraulic oil is pumped at very high pressures of about 200–300 bar.

This oil is fed on to the hydraulic motor which is mounted directly on the mill rollers, thus making redundant the entire speed-reduction gear system at the drive end. The hydraulic system avoids the waste of space and necessary maintenance of gears, and bearings and also avoids the energy loss which takes place at each speed-reduction stage. It also eliminates the use of ugly tail-bars, heavy muff-couplings, large crown pinions, etc.

Hydraulic motors are more easily controlled by remote, automated speed systems in addition to possessing all the advantages of steam turbines. They operate very quietly, unlike the turbines and gear trains.

An additional advantage is that hydraulic motors can be mounted on each roller, thus enabling each roller to be run at a slightly different speed.

The hydraulic motors being marketed are basically of two types:

1. Those with an output shaft speed of around 0–6 rpm (these are bulky and costly); they need to be mounted on each roller but do not involve any speed reduction gears.
2. Those designed to run at higher speeds of, say, 40–60 rpm but with an integral, enclosed speed-reduction device, such as a planatory gear box, to provide output speeds of 0–6 rpm. These are not so bulky in size and one motor can be used to drive an entire mill through the top roller with crown pinions to drive the other two rollers. These are relatively cheap, but the planatory gears may need some attention, although the hydraulic motor will not.

In both cases the mill-house width can be reduced from 24 to 18 m with a consequent reduction in the cost of buildings and drive foundations, use of smaller-span E.O.T (electrically operated travelling) cranes, more accurate automation, and energy savings.

DC Motors. With recent developments in solid-state electronics, it has been made easier to convert AC power to DC power and this, in turn, facilitates easier speed variation, automation, etc. Speed-reduction stages are also reduced.

AC power is generated using multi-stage turbines with a better thermal efficiency than the single-stage turbines generally used to drive mills. This results in a reduced use of steam for motive power generation.

This system is much cheaper than the hydraulic motor system, but involves a large number of speed-reduction stages and use of a tail-bar system, etc.

12.1.5.3 Free Rotating Discharge Rollers

A novel but simple adaptation to the conventional drive system was introduced by this author in 1982. The crown-pinion on the discharge roller was removed and this allowed the delivery roller to rotate due to frictional forces acting between the cane mat and the delivery roller. The frictional coefficient varies, depending on variations in fibre-loading; this results in corresponding variations in the speed of rotation of the delivery roller, ensuring that the cane mat is subjected to a uniform hydraulic loading irrespective of variations in the fibre load.

12.1.6 Mill Roller Grooves

Construction of a mill roller involves shrinking a cast iron shell on to a shaft. The shaft carries the load and is mounted on bearings at either end. The shell provides the crushing surface and can be renewed when it wears out.

The roller shell is shaped and machined to form different types of grooves with a view to providing a grip to hold the cane during juice drainage. The types of grooves in use are:

1. Radial V grooves.
2. Radial, narrow, deep grooves for juice drainage, generally known as *Maeshaert* grooves.
3. Longitudinal grooves in the form of chevrons to grip the cane.

12.1.6.1 Radial Grooves

The V grooves may be cut in different angles on the top, feed and discharge rollers, though the advantages claimed are not very justifiable, especially since roller arcing has come into vogue. Similarly, chevron grooves are being dispensed with as an adequate 'bite' is provided by the techniques of roller arcing.

Grooves, however, play a significant role in the effective performance of mills, though the details cannot yet be based on any pragmatic considerations.

12.1.6.2 Roller Arcing

For about two decades, the deposition of additional metal on the rollers in the form of globules, by adopting arc-welding techniques, has been universally adopted, though the areas and extent of such metal deposition (generally known as roller arcing) have not been standardized. Special machines and welding rods (even in a reel form) have been developed to enable controlled and continuous arcing even while crushing and juice flow is in progress.

Roller arcing has been a significant development in recent years and has contributed much to improving the capacity and efficiency of mills and to extending the life of roller shells.

12.1.7 Mill Miscellany

12.1.7.1 Hydraulic Loading

To apply the maximum possible pressure to the cane mat to extract the juice, the top roller, which provides the squeezing force at both the feed and discharge ends, is subjected to hydraulic pressures of as high an order as the geometry and metallurgy of the mill-house design and roller construction will permit. The pressure is applied through a hydraulic cylinder, a ram and pressure distribution plate on the top bearings at both ends. Usually, the pressure applied amounts to about 200 $t\,m^{-1}$ of roller length.

12.1.7.2 Mill Setting

Squeezing of the cane mat and expelling of the juice takes place at the feed and discharge ends of the mill. The relative openings, and also the hydraulic loading which results in some variation in such openings, have to be carefully adjusted, depending on the fibre content, rate of crushing, number of mills in tandem, extent of cane preparation, details of grooving, etc.

The main objectives to consider when determining the settings are:

1. Optimum extraction of juice.
2. Avoidance of chokes at the nip of the rollers at both ends.
3. Smooth travel of the cane mat over the stationary surface of trash plate.
4. Avoidance of 'slippage' of the cane during the squeezing process, resulting in reabsorption of juice etc.

This procedure has been fairly well rationalized, though it is still subject to the personal preferences of the engineers involved.

A simple, practical test to judge the effectiveness of the mill setting is to determine the brix of the feed-roller juice, discharge juice and the common mixed juice coming from the mill. This enables a calculation of the quantity of juice extracted at the feed and discharge ends, which should normally be in the ratio 3:1. The ratio of brix between the feed and discharge juices and the common juice should be 0.8 and 1.2 to indicate optimal imbibition efficiency. A wider range of brix values for feed and discharge juices would result only when imbibition water or the juice applied is extracted without effectively mixing. In some recent designs, like the FCB or Walkers, the top roller is fixed to a fulcrum whose geometry ensures that a constant ratio of feed: discharge openings is maintained. These amount to constant ratio mills, which is different from the self-setting mills that they claim to be.

12.1.7.3 Imbibition

As only about 80% of the juice present in the cane mat can be extracted in one mill (with two compressions), the only manner by which most juice (or more specifically, dissolved sugar) can be extracted is to replenish the extracted cane with diluted juice (or water at the end of the process) and repeatedly extract the juice in successive mills. The process of dilution and extraction is known as imbibition and is essentially to ensure that there is maximal extraction of juice or sugar.

The water is usually added at the point of entry to the last mill and then the juice extracted is pumped to the penultimate mill and so on, resulting in counter-current extraction, etc.

Hot water is obviously more helpful but at very high temperatures, say 90° C, there is a greater release of waxes and other non-sugar materials. Around 75–80 °C is found to be optimal.

The greater the amount of water used, the greater is the extraction, up to an approximate maximum of 400% water on fibre; above this the principle of diminishing returns applies. As more water is used, so the steam demands for evaporation also increase. Hence, as a good compromise, about 250–300% is used, although this depends very much on several other factors such as the extent of cell opening, the number of mills or compressions in the tandem hydraulic loading, the fibre loading, grooving, arcing and roller speeds.

In the diffusion process, the amount of water used is greater and the number of stages of such an application is also larger than in mills. It amounts to only dilution of juice. In the mills it is a displacement-cum-dilution, in a limited number of stages.

The LP (low pressure extraction) system is a compromise between the diffusion and HP (high pressure) milling system, i.e. an optimal combination of the larger stages of compression (juice displacement) and juice dilution.

The imbibition juices are recirculated using chokeless pumps. For more effective imbibition, without necessarily using larger quantities of water, many factories adopt a short circulation of the juices extracted by one mill on to the feed of the same mill, i.e. in addition to imbibition juice from the penultimate mill; this is generally known as integral imbibition. This increases the liquid: solid ratio, leading to a higher imbibition efficiency and a better leaching effect.

12.1.8 Juice Treatment and Clarification

The juice obtained from milling contains many impurities: floating solids, suspended colloidal matter and undesirable dissolved substances such as phenols, colouring compounds, starch, glucose, fructose, minerals, amino acids and other acids. These have to be eliminated as far as possible. The pH of the mill juice, which is around 5.6, has ultimately to be raised to 7+ so as to minimize the inversion of sucrose and destruction of reducing sugars.

The treatment generally consists of a combination of the following steps:

1. Initial heating of juices around 70–75 °C.
2. Addition of lime.
3. Addition of dissolved phosphates.
4. Bubbling of sulphur dioxide gas or carbon dioxide through the juice in one or more stages.
5. Second-stage heating to 100+ °C.
6. Addition of polyelectrolytes to aid coagulation of precipitates.
7. Sedimentation and decantation of the clear juice.
8. Filtration of the sediment to separate the solids and reprocessing of the juice portion.

These steps result in the production of clear juice (CJ) which is a transparent, light golden liquid that has lost most of its colloidals, gums, colouring compounds, etc., and whose pH has been adjusted to around 7+ . At this pH the clear juice can be concentrated without the least fear of either inversion or any other form of decomposition.

12.1.8.1 Juice Treatment Methods

The method used primarily depends on the nature of the end-product desired, though this will be sugar in every case. The grades of sugar produced are mainly (1) raw sugar (en route to the production of refined sugar) and (2) mill white (plantation white sugar).

Raw sugar, a product for shipping, needs to be redissolved in water and subjected to various refining treatments and crystallization operations. Raw sugar is not meant for direct consumption. It is produced in the form of crystals but these do not have a high degree of purity or dryness. They retain a thin coat of molasses and, hence, are a light yellow-brownish colour; they are not free flowing. Production of raw sugar in colony countries, and its refining in the western world or imperial countries is a continuation of the historical situation. Though colonies and imperial countries no longer exist politically, the economies developed in those times largely persist. Raw sugar is shipped to countries in Europe, America, Japan and lately the Middle East, where raw sugar is converted into refined white sugar for consumption.

12.1.8.2 Defecation

The clarification process adopted for raw sugar production is very simple and cheap. It consists primarily of an initial heating, the addition of lime and, in some cases, addition of a small quantity of phosphate.

Such a process is generally known as defecation. The treated juice is later heated to boiling point again before the decantation operation. For some juices which contain dissolved starch (which creates problems in the refineries), the clarified juice is subjected to an enzyme treatment to eliminate the starch.

12.1.8.3 Mill White Sugar

In many countries sugar is produced directly in the form of white, free-flowing crystals ready for direct consumption, i.e. without being redissolved and subjected to refining operations before the final crystallization. This is mostly done by adopting a clarification process involving the addition of lime (plus phosphate) and sulphur dioxide at two stages, i.e. at the juice stage and later after concentration, at the syrup stage. Hence, this is known as the double sulphitation process.

In some factories, carbon dioxide is used in place of sulphur dioxide and the process is then known as the carbonation process.

This direct processing yields a sugar of high quality which is comparable to refined sugar in its whiteness, crystal structure and sucrose content but which has traces of mineral salts, particularly $CaSO_3$. This becomes apparent when the sugar is dissolved in water. Refined sugar, with no ash content, produces a transparent solution while the sulphitation-produced sugar yields an opaque solution, the calcium salts becoming dispersed as a fine colloid.

12.1.8.4 Clarification Equipment

This consists primarily of:

1. Juice heaters which are mainly tubular heat exchangers using exhaust steam or vapour bled from one of the bodies of the evaporators.
2. A sulphitation tank which is generally a cylindrical tank with a conical bottom and gas distributors. The lime and sulphur dioxide are simultaneously and continuously let in. The equipment also includes a lime proportioner to regulate lime addition for proper clarification and settling. The pH is measured with a probe and the gas inlet is automatically regulated to ensure juice outflow at a predetermined pH.

   In the carbonation process carbon dioxide is used in place of sulphur dioxide but because of the nature of the precipitation, two stages are required. The carbonation process produces a better quality CJ and therefore a brighter sugar. But it is much more expensive and huge quantities of lime sludge are generated and create a waste problem. Hence, many factories are switching from the carbonation to the sulphitation process.

   With the double sulphitation process, several factories in India are producing sugar with an ICUMSA (International Commission for uniform methods of sugar analysis) value of 70–80, (refined sugar has a

ICUMSA value of about 50) which is very creditable. But the presence of opaque colloids in solution and even traces of sulphur make this sugar unsuitable for end-uses like beverages, food and fruit preservation in cans, pharmaceutical syrups, etc.

The equipment used in the carbonation process differs from that used in the sulphitation process only in detail.

3. Clarifiers. The multi-tray type of clarifiers, such as the Dorr clarifier, are commonly used. There has been no breakthrough in clarifier design and the present design has been in use for over 50 years; when changes are introduced, they are only in minor detail. In the defecation process, particularly in Australia, the trayless clarifier has been successfully replacing the multi-tray type, thus reducing the juice retention time from over 3 h to less than 1 h. This is indeed a breakthrough, but attempts to adopt it for sulphited juices have not been successful.
4. Filters. The old filter presses have been totally dispensed with and rotary drum vacuum filters are now universally adopted, both in the defecation and sulphitation systems. Here again, the design developed over 50 years ago continues to be in vogue and no new techniques have been introduced.

The weakness with the rotary vacuum filter is that it eliminates only part of the solids and the filtrate obtained is still very muddy. More than 20% of the solids, and the juices, need to be recirculated continuously.

Pressure-type filters, as used in filtering 'remelts' in the refineries, have still to be modified to filter the juice muds to produce a clear filtrate which can be mixed with the decanted CJ.

Both the clarifier and filter stations suffer from this lag in technology and require attention from the technologists.

12.1.9. Evaporation

The CJ drawn from the clarifier has a solid (brix) content of 13–15% and needs to be concentrated to a syrup with a brix of about 60%. This can be done in a multiple effect evaporator. Exhaust steam at a pressure of about 1–1.5 kg cm^{-2} and a temperature of 115–125 °C is used in the first body and a vacuum of about 650 mmHg (55 °C) is maintained in the last body. The number of effects are normally four (quadruple) or five (quintuplet).

In beet processing, an initial temperature of upto 150 °C and upto eight effects are in use. This is because beet juices contain no reducing sugars and consequently there is no fear of destroying the sugars resulting in the formation of coloured compounds. Hence, the beet industry can get away with using less steam than can the sugarcane industry.

12.1.9.1 Multiple Effect Evaporation: Vapour Bleeding

As stated above, the larger the number of effects, the greater the steam economy. But, the lesser the number of effects, the higher is the evaporation capacity of each effect. So, one has to find a compromise between the initial cost of the equipment and the extent of steam economy desired. However, there is a limit to the number of vessels that can be used because as the concentration increases, so a boiling point elevation (BPE) is reached which reduces the effective evaporation rate. Anything more than a five effect system is almost unworkable. Most factories install either a four-effect or a five-effect system for optimum economy.

Further economy can be achieved by withdrawing vapour from the different effects for use in pans or juice heaters. In recent installations using thin-film evaporators and exhaust steam at a temperature of about 125 °C, all the juice heating, as well as all the pan-boiling, is done with vapour. This is feasible because in the earlier bodies, when the juice is thin and has a lower BPE, the evaporation rates are higher.

A typical set-up for a 100-TCH plant using the double sulphitation process and producing white sugar with an average crystal size of 0.8 to 1.00 mm, would be as follows, presuming CJ of 100 TCH at 15 brix (total solids in solution) and syrup at 60 brix necessitates a total evaporation of 75 $t\,h^{-1}$.

Assumptions:
TCH: 100 (2400 t cane day^{-1})
Exhaust steam pressure: 1.2 $kg\,cm^{-2}$ (124 °C).

Vapour use:
Raw Juice heating: 7 $t\,h^{-1}$ (III body vapours)
Sulphur juice heating: 7 $t\,h^{-1}$ (I body vapour)
Pan boiling: 20 $t\,h^{-1}$ (II body vapours)

The assumed evaporation coefficients are conservative values and can easily be increased.

1.	Effect	I	II	III	IV	V
	Exhaust/vapour Inlet/outlet temp. (°C)	124/116	114/105	105/92	92/75	75/55
2.	Effective ΔT	8	10	12	15	16
3.	Juice inlet brix	15				
4.	Total evaporation ($t\,h^{-1}$)	35.5	28.5	8.5	1.5	1.5
					(Total 75.5 t)	
5.	Vapour use: Station and $t\,h^{-1}$	SJH 7	Pans 20	RJH 7	— —	— —
6.	Vapour to next body ($t\,h^{-1}$)	28.5	8.5	1.5	1.5	1.5
7.	Evaporation coefficient assumed ($kg\,m^{-2}\,1°C\,h^{-1}$)	4.5	3.0	2.00	1.0	1.0

8.	Heating surface required (rounded off) (m^2)	986 1000	950 1000	355 400	100 200	100 200
				(Total 2800 m^2)		

Exhaust required

Evaporator	35.5
Clear juice heater	2.5
	38.0% cane
Misc.	2.0
Total steam	40.0% cane

With the realization of the possibilities of extended steam economy by use of extensive and intensive vapour bleeding, it is no longer the fashion to provide equal heating surfaces (HS) in all the bodies. In recent installations the last body of a quintuple has an HS which is only 20% of the HS of the first body.

12.1.9.2 Evaporation Equipment

Conventional evaporators use a short-tube construction, generally known as the Roberts type. In these, the juice level in the tubes varies between 20 and 40% of the height of the tubes. With short tubes (2000 mm long) the calendria diameters are large and the juice retention time is long: 4–8 min. This gives scope for inversion and colour formation. Also, when the bodies are large, it is difficult to provide for proper distribution of the juice, steam, venting of non-condensable material, etc.

With sealed down-take the assumption is that the juice travels up the tubes only once and gets transferred to the next body through the down-take. But one can observe the juice bubbling up and down in the tubes, and, hence, juice distribution is bound to be irregular. Some designers have provided multiple down-takes in large bodies of conventional design.

Another disadvantage of the short-tube evaporator is the advice not to use exhaust steam at higher pressures and temperature. As the retention time is high, inversion is substantial and cannot be avoided. Consequently, the temperature differences (Δt) are limited, as are the evaporation coefficients. These disadvantages are overcome in the thin-film evaporators.

12.1.9.3 Thin-Film Long-Tube Evaporators

There are basically two different designs of long-tube, thin-juice-film evaporator and these are (1) the rising film evaporator (RFE) and (2) the falling film evaporator (FFE).

Rising Film Evaporators. These are similar to conventional short-tube evaporators, except that the tubes are 4 to 5 m long. Hence, for a given HS, the diameter of the body is considerably reduced.

The juice inlet is at the bottom. Due to the heat and ebullition, the juice travels up the tube rapidly and in the process becomes atomised. The mist, i.e. juice particles and vapour, travels upwards, resulting in wetting of the entire internal surface of the tube. Because the film is thin, the heat transfer coefficient is high. The mixture of juice and vapour is led from the top of the calendria into a flash chamber provided with very efficient mist separators, such that clean vapour without any carry-over of the juice is conveyed to the next body.

To ensure rapid boiling, it is essential that the juice entering the RFE is at around the boiling point of the juice in the body. Hence, the CJ emerging from the clarifier at around 95 °C needs to be heated to around 110–115 °C before it enters the RFE. In such a situation, the juice level in the calendria should be less than 15% of the tube length so that the BPE due to the hydrostatic head is reduced. Otherwise, if the temperature is well below the boiling point, the RFE must heat the juice. The liquid level inside the tubes then increases, as does also the BPE due to the increase in the hydrostatic head. This will allow inversion etc. to occur. Hence, with RFE, prior heating of the CJ to a temperature near the BP is essential.

Once the boiling point has been reached, the time taken for the thin film or mist to travel up the tube is very short, only about 20–30 s. In view of this, the total retention time, and hence the scope for inversion, is only a fraction of what it is in the short-tube evaporator. This, in turn, enables the use of higher temperature exhaust steam in the calendria. Though the higher temperature aids faster inversion, the effect is minimized because of the very short time for which the juice is at this temperature. Inversion, as with any other chemical reaction, also depends on a time factor.

The tendency to form a mist and rise in the tube is adversely affected by an increase in the brix (weight) of the solution. Hence, the use of an RFE should be restricted to the first one or two effects only; subsequent bodies should be of the short-tube type.

The heat transfer coefficient is higher, but then the evaporation coefficient depends on the extent of ΔT available and this can be enhanced only by extensive vapour bleeding for use in pans or juice heaters.

The RFEs are a great improvement on the short-tube evaporators but to get optimal results, greater care in operation is essential.

Falling Film Evaporators. These have made their entry into the sugarcane industry in very recent years. As the name indicates, the flow of juice is from the top to the bottom of the tube, i.e. the reverse of RFE. The juice enters at the top of the calendria and an ingenious design is necessary to ensure that the juice enters every tube, and does so continuously. As the juice travels down the tube, there is no need to preheat the CJ before it enters the FFE, though if this is not done, then a part of the HS of the FFE is required to serve as a heater. But, as in RFE, a lower temperature of the CJ does not create problems of hydrostatic head, increased inversion, etc.

As the juice trickles down the tube, so the vapours formed have a tendency to travel upwards. This causes turbulence which ensures the proper formation of a film and complete wetting of the internal surface. This is easy so long as the distribution arrangements are satisfactory and a continuous feed is ensured. For this, recirculation of the juice is invariably provided for.

Because the juice is falling, it travels very fast, and even if tube lengths of about 6–8 m are used, the duration of contact (retention) with the tube is in the order of 20–30 s only. With no scope for build-up of the level inside the tube, there is no further addition to the retention time. Hence, scope for retention of the juice is absent, except for the few seconds taken by the juice to travel down the tube.

Another advantage with FFE is that in view of its high heat transfer coefficient and freedom from any adverse effects of BPE due to either an increased concentration or the hydrostatic head, it is feasible to have six or seven effects compared with the limit of five effects now. FFE could be used for all the effects, including the last body. Undoubtedly, the cost of the equipment would be higher but as significant energy savings would be made in the years to come, the possibility of an increased number of effects proving beneficial is high, and the introduction of such a system is imminent.

The operation of FFE is simple and can be automated. Also juice feeding and removal are simple. On the other hand, scale removal is a bit more difficult but scale formation would be minimal. Chemical cleaning is easy to carry out.

12.1.10 Syrup Treatment

Syrup, as it emerges from the evaporators at around 60–65 Bx, is fairly dark in colour, and chemical treatment is necessary to remove the colour and reduce the viscosity.

12.1.10.1 Processing

In either of the processes involved in making mill white sugar, i.e. sulphitation and carbonation, the syrup is subjected to absorption of sulphur dioxide gas. This serves to both remove the colour and reduce the viscosity; it is also reported to contribute to a longer shelf-life of the sugar.

The operation is simple: the syrup is purged with SO_2 gas in a cylindrical tank, and the operation is continuous. This treatment results in a reduction of the pH to around 5.6.

The addition of a small amount of phosphoric acid also serves the purpose well, though it creates the minor problem of some sludge formation due to the precipitation of calcium salts.

Ion-exchange processes are also available with only brine solution to regenerate the resins; such a system has yet to prove popular.

12.1.10.2 Melts

The treatment that sugar melts are subjected to in the refinery is entirely different. Here, ion-exchange is the main method used, though chemical treatment using phosphate carbonation, or trickling the melt through beds of bone-charcoal or activated carbon, or a combination of these, is also generally used. The melt is also universally subjected to pressure filtration to eliminate any colloids etc.

It should be noted that before the preparation of the sugar melt commences, the raw sugar is mingled and subjected to centrifugal separation to rid it of the molasses. Continuous centrifugals are generally used and the melt should have a sucrose content of around 99%. Even so, the colloids, colouring compounds and dissolved minerals must be eliminated; hence the necessity for melt clarification despite its high purity. (Mill white sugar is boiled using a syrup of only 80–85% purity.)

12.1.11 Pan Boiling and Sugar Crystallization

This is the most important operation in the sugar factory, be it a refinery or a mill white unit.

The clarified syrup (melt) is further concentrated to a stage where sugar crystals form. This is again done under vacuum and at low boiling temperatures; however, the equipment, in design and construction, is very different from that of the evaporators where only syrup is formed and is not extended to the stage of sugar crystallization.

Skill, science and technology are involved in this processing stage: there is skill in creating crystals which are uniform in size, shape and colour; scientific precision is required to ensure that crystallization of only sucrose occurs and that crystallization and trapping of any colloidal or colouring compounds between crystal layers is avoided; and technology is involved in carrying out these operations fast and with due economy in the use of equipment, water, steam, energy and any other additives.

12.1.11.1 Crystallization

Crystallization and growth of crystals upto marketable standards are usually carried out in two or more stages, often three. These stages are dictated by the limitations posed by: the impracticability of crystallizing all the available sugar in syrup in one stage; the need to maintain crystal uniformity; and maintaining the nature of the massecuite such that the crystals and molasses separate efficiently during the centrifugal stage.

This has been made easy by the ability of sugar solutions to form supersaturated solutions so that the relative velocity of deposition of sucrose is higher than that of non-sugar material in the solution.

The syrup (melt) is steadily concentrated in the pans until the supersaturation stage is reached, when crystal nuclei form suddenly: this process may be further aided by techniques such as 'waiting' or 'shock-seeding'. The extent of nuclei formation must be carefully monitored so as to ensure their steady growth by not allowing fresh nuclei to be formed, nor allowing the nuclei already formed to dissolve again. This is achieved by the steady addition of syrup to maintain constant supersaturation and by regulating the rate of evaporation by controlling the admission of steam or vapour to the calendria.

Massecuite Systems. In practice, three grades of massecuite are produced, as indicated in Table 12.2.

Thus, all the commercial sugar is produced from A massecuite, while the final molasses is produced from the last massecuite, B, C or D depending upon the purity of the syrup, the superiority of the white sugar and the extent of the purity drop of the final massecuite.

Each massecuite is boiled so as to have a maximum crystal content, consistent with the fluidity of the massecuite for easy handling in centrifugals, and so as to provide the maximum exhaustion of sugar from its molasses.

Further crystallization of sugar from the final molasses is not possible by mere concentration and cooling, although it may still contain 30–35% sucrose. It is marketed for use as cattle feed or as a raw material for the production of ethyl alcohol by fermentation of the sugars present.

12.1.11.2 Vacuum Pans and Other Equipment

As the material being handled is a thick slurry with a high solid (crystal) content, the boiling equipment has to be specially designed to enable the

Table 12.2 Details of the three grades of massecuite

Massecuite	Apparent purity	Bx[a]	Product
A	80–90	90–94	Commercial sugar (white or raw); heavy molasses of 65–70 purity
B	65–75	92–96	Fine crystals to serve as seed for A; heavy molasses of 50–60 purity
C	50–55	98–102	Very fine crystals redissolved; final molasses of 28–32 purity

[a] Bx, hydrometer brix (total solids).

rapid circulation of this slurry; this is essential for the concentration and steady deposition of sugar on the existing crystals.

The heat transfer occurs through tubes in a calendria-type construction. The difference from, say, the evaporators, lies in the larger OD of the tubes (100 mm compared with 38 or 45 mm in the evaporators), their shorter length (900–1200 mm compared with 2000–8000 mm in the evaporators) and a very large down-take, approximately to 50% of the free cross section of the tubes. However, as pans are still mostly of the 'batch' type, they have to increase in height. As the height of massecuite above the top tube plate increases, so the circulation becomes sluggish. While the heating and vapour bubble formation in the tubes provide the motive force to aid circulation, the column of massecuite provides resistance; hence, for proper circulation, the tube height is not a hindrance but the massecuite height is. Similarly, the down-take area must be at least 50% of the total free area of all the tubes.

The total HS depends greatly on the temperature of the heating medium, or rather the temperature between the vapour in the calendria and that of the massecuite in the tube. As the height of massecuite is generally 1500 mm above the top tube plate, and the massecuite relative density is nearly 1.5, the BPE due to the hydrostatic head tends to be as high as 15 °C above the vapour temperature, which is usually around 55–58 °C; this greatly limits evaporation and can even cause dissolution of the crystallized sugar at the lower levels, thus undoing the work done earlier.

The circulation also depends on the massecuite and is most sluggish with final massecuites. This author prefers to provide different massecuite heights for different massecuites, 1500 mm for A, 1350 mm for B and 1200 mm for C, while using vapour at 105 °C in the calendrias.

12.1.11.3 Conti-Pans

These are the latest innovation in sugar factories. Though almost all other operations had been developed to be continuous in operation, pan-boiling had remained a batch operation. In recent years, several factories in different countries have installed continuously boiling pans and these may be used for all massecuites: A, B and C.

The conti-pans start with stored 'footings', prepared earlier in batch pans, of requisite crystal size, population and purity to suit the final massecuite, i.e. A, B or C. The syrup or molasses feeds are provided at various stages in the pan and the feed rates are all well regulated, either manually or by automated systems. The massecuite is discharged continuously from the other end and exits via a 'barometric seal' to a receiving crystallizer.

Different designers have adopted different types of calendrias and a different number of stages; there are calendrias with central down-take of a channel type or with peripheral down-take and calendrias with vertical tubes, as found in batch pans, or with horizontal tubes where the steam is inside the tube and the massecuite is outside. The number of stages varies from about 7 to 12 depending upon the overall crystal growth or intended purity drop of the molasses.

Mostly in conti-pans the massecuite height is only 300–800 mm above the tube plate and the down-take proportion is double what it is in the batch pans. As a consequence the HS:volume ratio is also different. As yet, however, there is no standard practice.

Conti-pan operation is simple, and the only difficulties experienced are: (1) incrustation of sugar on the internal pan walls, crevices and projections; and (2) in removing and storing the contents of the pan when it has been emptied for cleaning etc. The advantages of all continuous operations hold true, i.e. uniform flow of input and output, particularly steam or vapour consumption, uniformity of the product, less labour required, easy automation, etc.

12.1.11.4 Crystallizers

The massecuites when dropped, are still in a state of supersaturation and hence deposition of sucrose can be continued after massecuite is discharged. This is achieved by holding the discharged massecuite in a U-shaped receptacle with a sturdy stirring mechanism. Stirring in the crystallizers is a follow-up to circulation in the pans, only the mechanism is different.

As the degree of supersaturation is higher and the rate of crystallization is slower in lower purity massecuites, the time of retention of the massecuite in the crystallizer varies for A, B and C massecuites. To further aid effective crystallization, the stirrer mechanism includes provision for continuous water cooling. Recent practice has extended water cooling to all the massecuites, though the designed rate of cooling differs between them.

Conti-Crystallizers. The U-shaped batch crystallizers are nowadays converted to continuous operation by adopting the cascading technique. The cooling water is also made to travel in a counter-current manner to avoid the 'chilling' effect.

Recently, vertical crystallizers have been introduced for final massecuites; however, these have a serious defect. The stationary cooling elements (pipes) become coated with a thick layer of molasses, adversely affecting heat transfer. Their only advantage is space saving.

12.1.12 Centrifugals and Sugar Handling

After adequate cooling, the massecuite in the crystallizer is subjected to centrifugal separation of the crystals from the molasses. As the molasses is sticky, very high forces are required to separate it from the crystals. The operations and equipment used are similar regardless of whether the end-product is raw, mill white or refined sugar. Batch-type machines are used for commercial sugar, while continuous-type machines are used for other sugars.

12.1.12.1 Commercial Sugar

This is produced from A massecuite of high purity. As the product is to be free flowing, to give it a pleasing appearance, the crystals must be uniform. The centrifugal forces required are in the range 1000–1200 *g*. The centrifugals now in use are large and fully automated and include charging, washing, spinning (drying), ploughing and restarting steps. Capacities range from 500 to 1250 kg massecuite per charge and provide 16–20 cycles per hour. There is also scope to vary the speed depending on the purging quality of the massecuite.

White Sugar. The sugar is washed with superheated water and/or steam to totally eliminate all traces of molasses from the crystal surface and to reduce the moisture content to less than 2% when the sugar is discharged at a temperature of around 70–80 °C. Automatic separation of heavy and light molasses is provided for.

The sugar is discharged into a hopper type of conveyer and is taken to the sugar dryer.

Raw Sugar. When raw sugar is the desired product, then washing with water/steam is unnecessary. A thin film of molasses is required, as dictated by the regulations on international trade of raw sugars. However, it still needs to be dried.

12.1.12.2 Sugar Drying

In both raw sugar and refinery systems, the sugar discharged from the centrifugals is conveyed to the boot of a bucket elevator, from which it is discharged into the hopper which feeds the rotary drum sugar dryers. Hot air is blown in the counter-current direction. The dry sugar is then conveyed to silos for storage, prior to distribution.

With mill white sugar, popular in developing regions like central-western and southern Asia, Far East, Africa, etc., the market prefers uniform white crystal free flowing (dry) sugar but it must also have lustre and transparency. The sugar dried in rotary drums loses its lustre and develops an opaque quality, though it remains very white.

To achieve the desired end-product, drying in hoppers using hot air blown in a co-current direction is used.

12.1.12.3 Sugar Packing, Storage and Distribution

Traditionally, sugar, whether raw, white or refined, was bagged in burlap (jute) bags which were sometimes lined with thin cotton bags, particularly for refined sugar. Lately, however, packing patterns and systems have undergone a revolutionary change, and different systems are based on trading patterns as well as on consumer tastes.

International Trade. Raw sugar is mostly shipped loose in bulk, in ships with a large holding capacity. Hence, storage at the factory end is also in bulk, in large silos. Retrieval of stored sugar, loading into road or rail tankers, handling and loading onto ships are all carried out by totally mechanical and automated systems involving the loose sugar. The entire system is controlled by the weather so that the raw sugar, being hygroscopic, is not exposed to humid conditions even while being discharged into ship holds.

Mill white sugar is mostly packed in jute or polyethylene bags for storage or for distribution for domestic consumption. However, for international trade, it is packed in HDPE bags, weighing no more than 50 kg each. These are also used for refined sugar.

Consumer Packs In the developed world, the retail marketing of sugar has undergone a sea change. Sugar is now packed into small pouch packs, strip packs or is made into cube sugar packed in cartons.

For bulk consumers, like beverage or confectionery manufacturers, liquid sugar is transported in specially made bulk road tankers and it is stored, at both ends, in stainless steel tanks. Special grades of sugar, such as Demerara, brown sugar, etc., are marketed in consumer packs.

In most developing countries, however, factories sell sugar only in large bags, jute or HDPE, of 100 or 50 kg. Retailers then repack the sugar in smaller pouches. Table 12.3 gives details of the international standards set for the quality of the different types of sugar.

12.1.13 Bagasse Use; Steam and Power

Cane is rightly defined as energy cane; it can yield a surplus energy after meeting the captive needs of the sugar factory.

Sugar and fibre are both high calorie substances. Since sugar is extracted for use as a prime sweetening compound, molasses is used to 'add spirit' to

Table 12.3. International sugar quality standards

Parameter	Sugar grade		
	Mill white (CODEX)	Refined sugar	
		(CODEX)	(EEC)
Polarization (min.%)	99.5	99.7	99.7
Invert sugar (max.%)	0.1	0.04	0.04
Conductivity ash (max.%)	0.1	0.04	0.07
Sulphated ash	—	—	—
Moisture (%)	0.1	0.1	0.06
Colour (ICUMSA) (max.)	150	60	45
SO_2 (max. ppm)	70	20	—
Arsenic (max. ppm)	1	1	—
Copper (max. ppm)	2	2	—

the lives of many. This leaves only the bagasse and that is definitely not a waste product.

12.1.13.1 Bagasse Use

After the juice has been extracted cane yields about 30% bagasse (50% on a wet basis). This bagasse has a calorific value of around 2200 kcal kg^{-1}, even on a wet basis. This is a lot of energy and provides the sugar mills with enough to meet all its requirements for motive power and process steam as well as leaving something to spare.

As stated earlier, a modern mill white factory needs about 40% steam at a pressure of about 1.5 kg cm^{-2} (150 kPa). Bagasse with 50% moisture serves as a very efficient fuel and is used in the boilers at the factory to raise steam. Unlike in the early decades of this century, factories today install boilers with a working pressure of 42 kg cm^{-2} or more and the boilers are provided with efficient waste-heat recovery devices like economisers and air pre-heaters. With these, modern boilers operate at a thermal efficiency of about 69–70% on a GCV (gross calorific value) basis (compared with 55–60% previously) and are comparable with the best boilers using other forms of fuel. The bagasse/steam ratio is around 1:2.2.

12.1.13.2 Steam and Power Basis of Projections

The steam and power balance roughly work out as shown in Table 12.4.

Thus, a 100 TCH sugarcane mill can generate an income of US$ 6500 per day by sale of surplus power, after supplying the steam and power required

Table 12.4. Details of steam and power produced from bagasse

	Per hour	Per day
Cane (t)	100	2400
Energy required	2000 KWh	3.2 MW at 0.8 pf (Power factor)
Steam required at 1.5 kg/cm^{-2} (t)	40	960
Bagasse used (t)	18	432
Power generation at 7.5 Kg Stm./KW	5.3 MWh	5.3 MW
Power used at 1.3 *k*W/TCD	3.2 MW	3.2 MW
Surplus bagasse (t)	12	288
Surplus steam potential (t)	26.4	633
Power Potential in condensing turbines at 5.5 kg/*k*W	4.8 MWh	4.8 MW
Total power potential	10 MW	10 MW
Power used-captive	3.2 MW	130,000 KWh
Extra income by power sale at US cents 5 per *k*Wh	—	US$6500

Figures have been rounded off for simplification

by the sugar mill. The scope can be further improved by installing boilers with higher working pressures of, say, 60 or 100 kg per cm^2., instead of 42 kg assumed.

There is also scope for further reduction in steam use at the sugar mill, by the installation of FFE evaporators, conti-pans and vapor recompressors, and by the adoption of membrane technology (reverse osmosis, etc.,) and by drying bagasse. The steam thus saved, can generate more surplus power.

12.1.13.3 Equipment

Boilers As stated earlier, today's boilers have high thermal efficiencies due to:

1. Improved furnaces of bagasse-feeding devices (spreader stokers, fluidized bed furnaces, etc.).
2. Automatic draught maintenance.
3. Automatic water supply and water level maintenance equipment and installation of heat-recovery devices, e.g. economisers and air heaters. Thus, there is a reduction in the flue gas temperature to 150 °C compared with 240 °C used previously.
4. Adoption of higher steam pressures and temperatures, etc.

Turbo-Generators. Multi-stage steam turbines produce a lot of power while using steam at higher pressures and temperatures. Surplus steam can be used in condensing turbines to generate yet more power. The power generators used generate power at higher voltages, i.e. 11 kV compared with LT (low tension) 440 V previously.

These compact units are easy to operate and maintain and can be synchronized with the public grid distribution system.

Automation and Control. The present turbo-generators are very sophisticated with electronic digital controls. They can be programmed to provide specific conditions, say maintaining back-pressure steam at a predetermined level, or producing power chiefly for the factory with only the surplus being supplied to the grid or vice versa.

All the motive power units in the sugar mills are driven electrically. With solid-state technology, the DC motor drives used in mills are very efficient and economical. Another interesting development is the introduction of hydraulic motors for mill drives; these avoid the long train of speed-reduction devices and associated loss of energy. DC motors and hydraulic motors are also cleaner, quieter and precisely automated.

The sugar industry is now reaching new levels of efficiency and automation, similar to those seen in industries like petro-chemicals, metallurgy, man-made fibres, etc.

12.2 By-Products

A schemetic diagram of by-products from the sugarcane industry is presented in Fig. 12.2 (Paturau 1989). In addition to sugar, about 38 products of commercial importance are produced with a value of approximately US $8000 per ton of raw material (Paturau 1986). High-value by-products of the cane industry include monosodium glutamate from molasses, furfural and cellulose from bagasse and lysine, industrial alcohol and yeast.

In India, *khandsari* is an old process of white sugar manufacture. It is an open pan process (in contrast to the vacuum process of the white sugar industry) and *khandsari* sugar is yellow to whitish in colour, opaque and dull in texture. It is either powdery or very small grained. It is supposed to be easily digestible with improved nutritive qualities. The modern *khandsari* units use an open-pan and lime-sulphitation process to improve the recovery and quality of sugar and the product resembles the white crystalline sugar

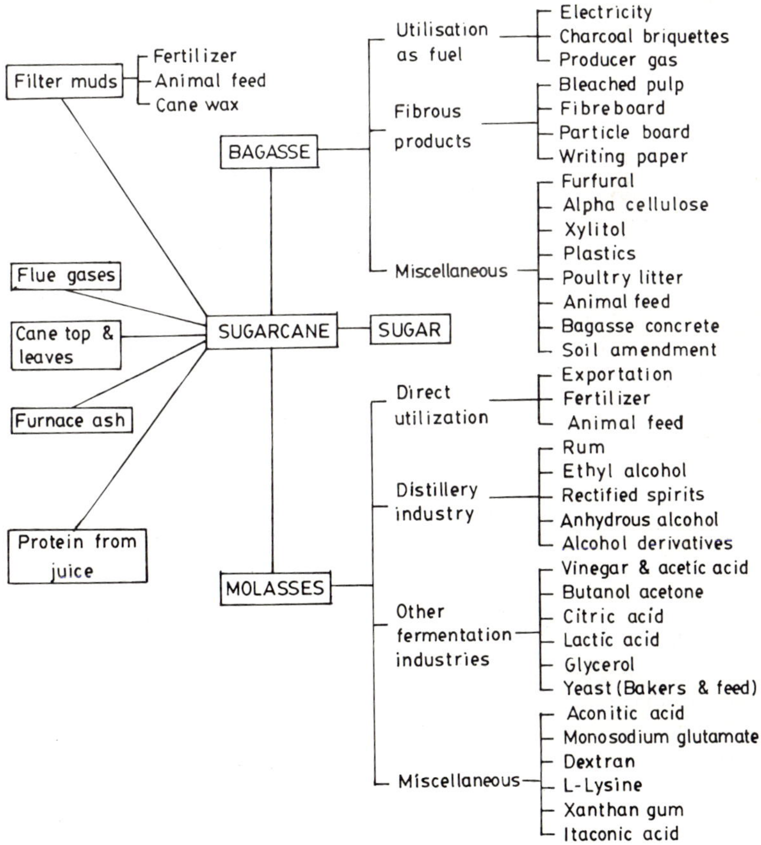

Fig. 12.2. By-products from the Sugarcane industry

produced by the vacuum pan industry. Sugar recovery from the *khandsari* method is around 5–9%. The units have a hydraulic crushing unit (providing about 70% extraction), a sulphitation tank, wet bagasse burning equipment and a horseshoe furnace.

Jaggery is another sweetening agent used in many parts of Asia, Africa and Latin America. In India jaggery manufacture is well standardized. It is an amorphous and crude form of raw sugar processed in an open pan, and the cane variety plays an important part in its quality and recovery of jaggery. Suitable sugarcane varieties for this purpose are CO997, CO527, CO421, KHS2045 (non-arrowing) and CO62175. Good jaggery is a light golden yellow and hard in consistency. The steps involved in its manufacture are:

1. Extraction of juice.
2. Purification.
3. Concentration.

The normal levels of juice extraction are around 55 to 60%. A six-roller hydraulic crusher with a crushing capacity of 4 $t\,h^{-1}$ is supposed to extract 80% and give a jaggery recovery of 10–12%. The extracted juice is placed in wide, shallow, iron pans (210–270 cm diameter and 45 cm deep). The pH of the juice is acidic, i.e. 5.8–6.8, and chemical clarificants like lime are added to neutralize the pH and facilitate flocculation of colloidal matter. Lime sucrate (1 unit of lime in 5 units of water) is another clarificant used, and other chemicals which are commonly used are sodium carbonate, super phosphate, alum and sodium hydrosulphate. In addition to these, vegetable clarificants are also used; these include the bark of *Kydia calycina*, seeds of groundnut and caster and, most often, the mucilage from the entire plant of *Hibiscus faculneus* and *H. escaluntus*.

Table 12.5. Chemical analysis of good (mature cane) and poor (immature cane) jaggery (cv. CO419). (Lakshmikantham 1983)

	Bad jaggery	Good jaggery
Moisture (%)	4.57	7.48
Glucose (%)	11.31	8.09
Sucrose (%)	76.43	81.09
Ash (%)	1.97	1.40
Total non-sugars (%)	8.04	3.61
Organic non-sugar (%)	6.07	2.21
Total N (%)	0.05	0.04
Total P_2O_5 (%)	0.1047	0.1697
Lime CaO (%)	0.1568	0.1101
pH	6.2	5.6
Colour	Dark yellow	Light golden-yellow
Hardness and crystallinity	Hard and crystalline	Hard and crystalline

Initially the juice is boiled at a low temperature with an appropriate quantity of the chemical clarificant, like lime or alum. This will facilitate flocculation of the colloids and the scum is skimmed off. At this stage extracts of vegetable clarificants are added and the magma is further purified and boiled at temperatures between 118 and 124 °C. At the striking point (i.e. 120 °C) the magma will form a long silky thread when the wooden ladle is taken out of the pan and held in the air. At this stage boiling is nearly completed. The pan is tilted in a cooling pan and different jaggery moulds are made. The storage of jaggery is also important for it must stand for 1 year; it is usually stored in gunnybags in elevated places. One study has shown that a good jaggery can be obtained only from a mature cane and this stores well for a longer period (Table 12.5).

Cream jaggery is manufactured with the use of activated carbon prepared from paddy husk. After removal of the initial scum, activated carbon at 1% of juice weight is added to the warm juice and this is heated again for some time. The juice is then passed through a bed of activated carbon and sand. The juice is practically colourless or light yellow. When this juice is boiled further with a small quantity of milk, added as a clarificant, cream jaggery is obtained. This is a light golden-yellow and has an excellent consistency (Lakshmikantham 1983). Cream jaggery used to attract a very high premium but now it is of academic interest only. Presently, however, it has a great export potential with a marked change in the export policy of the government of India.

References

Abayomi AY (1987) Growth, yield and crop quality performance of sugarcane cultivar CO957 under different rates of application of nitrogen and potassium fertilizers. J Agric Sci Camb 109: 285–292

Abayomi AY, Etejere EO, Fadayomi O (1988) Effect of N levels and water deficit on nitrate reductase activity (NRA) from leaves of two sugarcane cultivars: CO957 and CO1001. Turrialba 38: 352–358

Abbot EV (1959) Introduction to yield decline. Int Soc Sugarcane Technol Proc 10: 51–60

Abruna-Rodriguez F, Vicente-Chandler J (1967) Sugarcane yields as related to acidity of a humid tropic Ultisol. Agron J 59: 330–332

Adams RM, Rosenzweig C, Peart RM, Ritchie JT, McCarl BA, Glyer JD, Curry BC, Jones JW, Boote KJ, Allen Jr HL (1990) Global climate change and US agriculture. Nature (Lond) 345: 219–224

Addiscott TM (1970) The uptake of initially available soil potassium by rye grass. J Agric Sci Camb 74: 123–129

Agarwala SC, Chatterjee C, Nautiyal BD, Dube BK, Nautiyal M (1986) Induction of deficiency of zinc, copper and molybdenum in sugarcane and its effect on growth, sugar concentration and some enzyme activities. Sugarcane (UK) 1: 10–17

Agarwala SC, Sharma CP, Srivastava SC, Chatterjee C (1987) Boron nutrition of sugarcane. Sugarcane (UK) 5: 1–12

Agnihotri VP (1983) Diseases of sugarcane. IBH New Delhi 363 pp.

Alexander AG (1973) Sugarcane physiology: a comprehensive study of the *Saccharum* source to sink system. Elsevier, Amsterdam, 752 pp.

Alexander AG (1985) The energy cane alternative. Sugar Series 6. Elsevier, Amsterdam, 509 pp

Alexander AG (1990) High growth sugarcane as a natural conservation system for crop land, atmosphere and solar radiation. XVI Symp Natural resources, Sept 13, 1990, Puerto Rico, 29 pp

Aloma J, Perez H, Cuellar I (1974) Advances in sugarcane fertilization in Cuba. Int Soc Sugarcane Technol Proc 15: 608–617

Anderson DL, Snyder GH, Martin FG (1991) Multiyear response of sugarcane to calcium silicate slag of Everglades Histisols. Agron J 83: 870–874

Anonymous (1970) Annual report. Bureau of Sugar Exp Stn (B.S.E.S): 21

Anonymous (1980) Annual report. Bureau of Sugar Exp Stn (B.S.E.S): 18

Anonymous (1981) Annual report. Indian Institute of Sugarcane Research, Lucknow 1981: 83–84

Anonymous (1986) Exploitation of Indo-American (IA) clones. Sugarcane Breed Inst Newsl 5: 2

Anonymous (1987a) Annual report. Bureau of Sugar Exp Stn (B.S.E.S) 1970 (p21), 1980 (p18), 1987 (p16)

Anonymous (1987b) Annual report. Bureau of Sugar Exp Stn (B.S.E.S) 87th: 12–17

Anonymous (1988) Annual report. South African Assoc Exp Stn 1987–88

Anonymous (1990a) Frontiers in plant biology. Science 250: 881

Anonymous (1990b) Sugar statistics. Indian Sugar XL, Ann Sept 1990: 492–493

Arnold PW, Tunney H, Hunter F (1968) Potassium status: soil measurements and crop performance. Trans 9th Int Congr 1 Soil Sci Adelaide: 613–620

Artschwager E, Brandes EW (1958) Sugarcane. Agricultural Hand Book 122. US Govt Printing Office, Washington DC, 307 pp

Arzola PN, Carrandi JA (1982) The effect of filter cake application dissolved in water on cane sugar production. Entro Agrico (Cuba) 9: 89–108

Atchison JE (1976) Agricultural residues and other non-wood plant fibres. Science 191: 768–772

Atchison JE (1982) Utilisation of unconventional raw materials. IPPTA seminar held at Lucknow, UP India, December 1982. 61 pp

Atchison JE (1983) Bagasse newsprint. Has it arrived? Past history, present status and future possibilities. Int Soc Sugarcane Technol Proc 18: 1436–1461

Atchison JE (1986) Bagasse newsprint. The worldwide impact of commercial success after 130 years of efforts strewn with failures. Keynote address. International seminar on bagasse newsprint, Madras, India, pp 1–56

Austin RB (1982) Crop characteristics and the potential yield of wheat. J Agric Sci Camb 98: 447–453

Ayres AS (1959) The growth of sugarcane as influenced by phosphorus. The critical range of soil phosphorus. Int Soc Sugarcane Technol. Proc 10: 462–467

Ayres AS (1966) Calcium silicate slag as a growth stimulant for sugarcane on low- silicon soils. Soil Sci 101: 216–277

Ayres AS, Hagihara HH (1955) Soil analysis as indexes to nutrient availability. Hawaii Plant Rec 113: 120–126

Ayres AS, Hagihara HH, Stanford G (1965) Significance of extractable aluminium in Hawaiian sugarcane soils. Soil Sci Soc Am Proc 29: 387–392

Bajwa MI, Qidwai A (1986) Role of complex fertilizers in increasing fertilizer use efficiency: Pakistan experience. In: IFA Regional Agricultural Meeting Proc, Dec 8–12, 1986, ICAR, New Delhi, pp 63–68

Baker KM (1991) Biotechnology – a strategic opportunity for agriculture. Outlook Agric 20: 79–82

Banath LL, Monteith MH (1966) Soil oxygen deficiency and sugarcane root growth. Plant Soil 25: 143–149

Barnes AC (1974) Sugarcane. Leonard Hill Books, London, 572 pp

Batchelor CH, Soopramanien GC, Nayamuth R (1989) Design and management of sugarcane drip irrigation system. Int Soc Sugarcane Technol Proc 20, 2: 522–531

Batchelor CH, Soopramanien GC, Bell JP, Nayamuth R, Hodnett MG (1990) Importance of irrigation regime, dripline placement and row spacing in the drip irrigation of sugarcane. Agric Water Manage 17: 75–94

Baver LD (1956) Soil physics. Wiley, New York, 489 pp

Beaufils ER (1972) Diagnosis and recommendation integrated system (DRIS): a general scheme for experimentation and calibration. Soil Sci Bull 1 University of Natal, South Africa

Beckett PHT (1971) Potassium potentials—a review. Potash Rev 5(30): 1–41

Beckett PHT (1972) Critical cation activity ratios. Adv Agron 24: 379–411

Bernstein L, Francis LE, Clark RA (1966) Salt tolerance of NCO varieties of sugarcane. I. Sprouting, growth and yield. Agron J 58: 489–493

Biswas BC (1991) Magnesium status in China. Lessons for India. Fert News 36: 41–47

Blackburn F (1984) Sugarcane. Longman, New York, 388 pp

Blaney HF, Criddle WD (1950) Determining water requirements in irrigated areas from climatological and irrigation data. US Dept Agric (SCS) TP-96, 48 pp

Boddey RM, Urquiaga S, Dobereiner J (1991) Biological nitrogen fixation associated with sugarcane. Plant Sci 137: 111–117

Borden RJ (1948) Nitrogen effects upon the yield and composition of sugarcane. Hawaii Plant Rec 52: 1–51

Bowen JE (1972) Essentiality of chlorine for optimum growth of sugarcane. Int Soc Sugarcane Technol Proc 14: 1102–1112

Bowen JE (1981) Micro-element nutrition of sugarcane. 1. Effect of micro-elements on growth and yield. Trop Agric (Trinidad) 58: 13–21

Bowen JE (1983) Micro-element nutrition of sugarcane. 3. Critical micro-element levels in immature leaf sheaths. Trop Agric (Trinidad) 60: 133–138

Boyer JS (1982) Plant productivity and environment. Science 218: 443–448

Brady NC (1984) The nature and properties of soils. Macmillan, New York, 750 pp

Brandes EW, Sartoris GB (1936) Sugarcane: its origin and improvement. USDA Year B Agric 1936: 561–611

Bray RH (1954) A nutrient mobility concept of soil–plant relationship. Soil Sci 78: 9–12

Bray RH (1965) Modern trends in soil fertility assessment. Adv Agric Sci and their applications. In: Krishnamurthy S (ed) Golden jubilee of Madras Agric J, pp 269–278

Brodie HW, Yoshida R, Nickell CG (1969) Effect of air and soil temperature on growth of sugarcane clones. Hawaii Plant Rec. 58: 21–52

Brown RH (1978) A difference in N use efficiency in C_3 and C_4 plants and its implications in adaptation and evolution. Crop Sci 18: 93–98

Brown RH (1985) Growth of C_3 and C_4 grasses under low N levels. Crop Sci 25: 954–957

Brzesowsky NJ (1986) Factors affecting sugarcane production and quality. Agri Int 38: 30–34, 91–94

Bull TA (1969) Photosynthetic efficiencies and photorespiration in Calvin cycle and C_4 dicarboxylic acid plants. Crop Sci 9: 726–729

Bull TA (1975) Row spacing and potential productivity in sugarcane. Agron J 67: 421–423

Bull TA, Glasziou KT (1963) The evolutionary significance of sugar accumulation in *Saccharum*. Aust J Biol Sci 16: 737–742

Bull TA, Glasziou KT (1975) Sugarcane. In: Evans LT (ed) Crop physiology. Cambridge Univ Press, London, pp 51–72

Buren L (1976) Sugarcane response to the chemical ripener Polaris. Proc sugarcane ripener seminar, Orlando, Florida, May 2–4, 1976. Monsanto Agric Product Co. St Louis, Missouri, pp 84–93

Buringh P, VanHeemst HDJ, Staring GJ (1975) Computation of the absolute maximum food production. Agricultural University, Wageningen, Netherlands, 277 pp

Burr GO (1969) A note on the efficiency of sugarcane. Taiwan Sugar Exp Stn Rep 3: 1–4

Burr GO, Hartt CE, Brodie HW, Tanimoto T, Kortschak HP, Takahashi D, Ashton FM, Coleman RE (1957) The sugarcane plant. Annu Rev Plant Physiol 8: 275–308

Camp CR, Carter CE (1983) Sugarcane yield response to subsurface drainage for an alluvial soil. Trans ASAE 26: 1112–1116

Campbell RB (1967) Sugarcane. In: Hagan RM, Haise HR, Edminster JW (eds) Irrigation of agricultural lands. ASA Monograph 11, Wisconsin, USA, pp 649–654

Capo BG (1965) A new fertilizer yield equation. Int Soc Sugarcane Technol Proc 12: 380–389

Capo BG (1974) Optimum economic quantity of fertilizer based on plant tissue composition. In: Fernandez HV (ed) Fertilizer, crop quality and economy. Elsevier, Amsterdam, pp 45–71

Carter CE, Irvine JE, McDaniel V, Duncleman J (1985) Yield response of sugarcane to stalk density and subsurface drainage treatment. Trans ASAE 28: 172–178

Carter CE, Fouss JL, McDaniel V (1987) Water management increases crop yields. Pap Am Soc Agric Eng 87-2049, 11 pp

Cavalecante Va, Dobereiner J (1988) A new acid-tolerant nitrogen-fixing bacteria associated with sugarcane. Plant Soil 108: 23–31

Chang Hao, Wang JS (1968) Evapotranspiration and water requirement of sugarcane in Taiwan, Int Soc Sugarcane Technol Proc 13: 664–670

Chang Hao, Wang JS, Ho FW (1968) The effect of different pan ratios for controlling irrigation of sugarcane in Taiwan. Int Soc Sugarcane Technol Proc 13: 652–663

Chang HC, Chang CH, Ho FW (1969) Competition between sugarcane and intercrops for fertilizer tagged with ^{32}P and ^{86}Rb. J Agric Assoc China 63: 43–49

Chang Jen-Hu (1961) Microclimate of sugarcane. Hawaii Plant Rec 56: 195–225

Chang Jen-Hu, Cambell RB, Brodie HW, Baver LD (1965) Evapotranspiration research at the HSPA experimental station. Int Soc Sugarcane Technol Proc 12: 10–24

Chapman LS (1985) A review of sulphur nutrition of sugarcane in Queensland. Proc Aust Soc Sugarcane Technol (unnumbered) Proc of 1985 Conf, Bur Sugar Exp Stn, Mackay, Queensland, pp 113–120

Chapman LS, Hogarth DM, Leverington RC (1983) Does N fertilizer carry over to succeeding crops. Aust Soc Sugarcane Technol Conf April 18–22, 1983, Mackay, Queensland, pp 106–114

Chen Y, Barak P (1982) Iron nutrition of plants in calcareous soils. Adv Agron 35: 217–240
Chow P, Samuels G (1977) Evaluation of N fertilizer efficiency for plant and ratoon crops in irrigated sugarcane. Int Soc Sugarcane Technol Proc 16: 1259–1264
Clements HF (1940) Integration of climatic and physiologic factors with reference to the production of sugarcane. Hawaii Plant Rec 44: 201–223
Clements HF (1959) Quality in sugar production. Field aspects. Proc Hawaii Sugar Technol Assoc 18: 16–24
Clements HF (1964) Interaction of factors affecting yield. Annu Rev Plant Physiol 15: 409–442
Clements HF (1965) The roles of calcium silicate slag in sugarcane growth. Rep Hawaii Sugar Technol Proc: 103–126
Clements HF (1980) Sugarcane crop logging and crop control: principles and practices. The University Press of Hawaii, Honolulu, 520 pp
Clements HF, Nakata S (1965) Minimum temperatures for sugarcane germination. Int Soc Sugarcane Technol Proc 12: 554–560
Clements HF, Shigeura G, Akamine EK (1952) Factors affecting the growth of sugarcane. Univ of Hawaii Agric Exp Stn Tech Bull 18: 90 pp
Clowes M St J (1980) Ripening activity of the glyphosate salts Mon 8000 and Round up. Int Soc Sugarcane Technol Proc. 17: 676–693
Coleman RE (1968) Physiology of flowering. Int Soc Sugarcane Technol Proc 13: 992–1000
Coleti JT, Lorenzetti JM, Garla JH, Campponez A (1986) The inhibition of flowering by ethephon and its influence on sugarcane quality in Brazil. Int Soc Sugar Tech Proc Abstr Pap 19: 63
Criddle WD, Kalisvaart C (1967) Sub irrigation. In: Hagan RM, Haise HR, Edminster TW (eds) Irrigation of agricultural lands. ASA 11, pp 905–921
Das UK (1933) How to measure effective temperature in terms of day-degrees. Hawaii Plant Rec 37: 174–178
David H (1986) The internode borer *Chilo sacchariphagus indicus* Kapur. In: David H, Eswaramoorthy S, Jayanthi R (eds) Sugarcane entomology in India. Sugarcane Breeding Institute, Coimbatore, pp 121–134
David H, Nandagopal V (1986) Pests of sugarcane – distribution, symptomatology of attack and identification. In: David H, Eswaramoorthy S, Jayanthi R (eds) Sugarcane entomology in India. Sugarcane Breeding Institute, Coimbatore, pp 1–29
Davies T (1990) Grasses more sensitive to SO_2 pollution in conditions of low irradiance and short days. Nature (Lond) 284: 483–484
Davison AW, Bailey IF (1982) SO_2 pollution reduces the freezing resistance of rye grass. Nature (Lond) 297: 400–402
Deacon HFE (1968) Experience in the use and development of plantation-scale fully mechanised harvesting in the Caribbean. Int Soc Sugarcane Technol Proc 13: 1496–1512
Deerr N (1948) Sugarcane in legendary India. Indian sugar supplement. The Indian Sugar Syndicate, Kanpur, India, pp 7–8
Dent D (1986) Acid sulphate soils: a base line for research and development. ILRI Publ 39, Wageningen, The Netherlands, 204 pp
Dobereiner J (1961) Nitrogen fixing bacteria of the genus *Beijerinckia* Derx in the rhizosphere of sugarcane. Plant Soil 25: 211–216
Dodsworth GH, Nixon DJ, Sweet CPM (1990) An assessment of drip irrigation of sugarcane on poorly structured soils in Swaziland. Agric Water Manage 17: 325–355
Donald CM (1968) The breeding of crop ideotypes. Euphytica 17: 385–403
Donaldson RA, Van Staden J (1989) A review of chemicals used as ripeners of sugarcane in South Africa. Int Soc Sugarcane Technol Proc 20, 2: 647–655
Driessen PM (1986) Nutrient demand and fertilizer requirements. In: Van Keulen H, Wolf J (eds) Modelling of agricultural production: weather, soils and crops. Wageningen, The Netherlands, pp 182–202
Duflo D (1976) Some results of potassic fertilization trials on sugarcane in Africa. Potassium in soils, crops and fertilizers. Bull 10 Indian Soc Soil Sci, New Delhi, pp 286–292
du Toit JL (1959) Recent advances in nutrition of sugarcane in South Africa. Int Soc Sugarcane Technol Proc 10: 432–441

du Toit JL (1968) The basis of a fertilizer advisory service. Int Soc Sugarcane Technol Proc 13: 800–808

Dutt AK (1962) Sulphur deficiency in sugarcane. Emp J Exp Agric 30: 257–262

Elawad SH, Gascho GJ, Street JJ (1982a) Response of sugarcane to silicate sources and rate. I. Growth and yield. Agron J 74: 481–484

Elawad SH, Gascho GJ, Street JJ (1982b) Response of sugarcane to silicate sources and rate. II. Leaf freckling and nutrient content. Agron J 74: 484–486

El Messaoudi M (1990) Use of pan evaporation for estimating the total dose and programming the irrigation of sugarcane. Agric Water Manage 17: 209

Elwali AMO, Gascho GJ (1984) Soil testing, foliar analysis and DRIS as guides for sugarcane fertilization. Agron J 76: 466–470

Elwali AMO, Le Grand F, Gascho GJ (1980) Nitrogen leaching from soil and uptake by sugarcane from various urea-based fertilizers. Soil Sci Soc Am Proc 44: 119–122

Enahari JH, Yung Sheung RMG, Madrane A, Hesse FRW (1989) Intercropping sugarcane in Morocco. Int Soc Sugarcane Technol Proc 20, 2: 545–550

Evans H (1935) The root system of the sugarcane. I. Methods of study. Emp J Exp Agric 3: 351–362

Evans H (1959) Elements other than nitrogen, potassium and phosphorus in the mineral nutrition of sugarcane. Int Soc Sugarcane Technol Proc 10: 473–508

Evans H (1965) Toxic sulphate soils in British Guiana and their improvement for sugarcane cultivation. Int Soc Sugarcane Technol Proc 12: 695–714

Ewart GY (1965) Defining osmotic pressure in salinity problems. Int Soc Sugarcane Technol Proc 12: 82–89

FAO (1974) Food and Agricultural Organisation of the United Nations. production year book 28: 118

FAO (1988) Food and Agricultural Organisation of the United Nations, production year book 42: 207–208

Faris JA (1926) Cold chlorosis of sugarcane. Phytopathology 16: 885–891

Fernandes J, Furlaninejo VL, Stolf R (1981) Seed bed preparation of planting sugarcane and its effect on ratoon crops. Brs Acucareiro 97(4): 37–40 (Abst Trop Agric 7, June 1981: 123)

Fernandez Caldas E, Moreno HJ, Perez BA, (1975) The Q/I relationship for potassium in soils of different types of the island of Tenerife. Potash Rev 4(58): 1–8

Fogliata FA, Aso PJ (1965) The effects of soil-soluble salts in sucrose yield of sugarcane. Int Soc Sugarcane Technol Proc 12: 682–694

Fonseca JR, Egana J (1989) Irrigation systems and water use by sugarcane in Cuba. Int Soc Sugarcane Technol Proc 20: 514–520

Fox RL (1976) Sulphur and nitrogen requirements of sugarcane. Agron J 68: 891–896

Fox RL (1983) Principles of phosphate fertilization applied to sugarcane. Hawaiian Sugar Technologists' 42nd Annu Conf HSPA, Honolulu, pp 86–91

Fox RL (1986) Phosphorus – a basic nutrient for soil improvement. In: Pretty KM, Dowde SF (eds) Proceedings of the international conference on the management and fertilization of upland soils in tropics and subtropics. Institute of Soil Science, Nanjing, China, pp 57–63

Fox RL, Silva JA, Plucknett DL, Teranishi DY (1969) Soluble and total silicon in sugarcane. Plant Soil 30: 81–92

Fox RL, Bosshart RP, Sompongse D, Lin Mu-Lien (1990) Phosphorus requirements and management of sugarcane, pineapple and banana. In: phosphorus requirements of sustainable agriculture in Asia and Oceania. Proc Symp, 6–10 March, 1989. Int Rice Res Inst, Manila, pp 409–425

Francis A (1989) Biological efficiencies in a multiple cropping system. Adv Agron 42: 3–29

Freeman KC (1968) Influence of row spacing on yield and quality of sugarcane in Georgia. Agron J 60: 424–425

Gallagher JN, Biscoe PV (1978) Radiation absorption, growth and yield of cereals. J Agric Sci Camb 91: 47–60

Garza AM (1968) Aspects of climate on sugarcane. Int Soc Sugarcane Technol Proc 13: 832–836

Gascho GJ, Shih SF (1981) Cultural methods to increase sucrose and yield of sugarcane. Agron J 75: 999–1003

Gascho GJ, Anderson DL, Ozaki HY (1986) Cultivar-dependent sugarcane response to nitrogen. Agron J 78: 1064–1069
Gaunt JK (1965) Consideration in the mechanical harvesting of sugarcane. Int Soc Sugarcane Technol Proc 12: 354–363
Gayle GA, Skagogs RW, Carter CE (1987) Effects of excessive soil water conditions on sugarcane yield. Trans ASAE 30: 993–997
Ghosh AK, Rai RK, Saxena YR, Shrivastava AK (1990) Effect of sulphur application on the nutritional status, yield and quality of sugarcane. J Indian Soc Soil Sci 38: 73–76
Glaz B, Ullao MF, Parrado R (1989) Cultivation, cultivar and crop age effects on sugarcane. Agron J 81: 163–167
Govinden N, Arnason JT (1990) The relative importance of competition for water and for light in intercropping of sugarcane with maize. Agric Water Manage 17: 233
Grassl CO (1946) Cytogenetics of modern sugarcane. Econ Bot 17: 97–105
Grassl CO (1969) *Saccharum*. Names and their interpretation. Int Soc Sugarcane Technol Proc 13: 868–875
Grimme H (1976) Soil factors of potassium availability. In: Potassium in soils, crops and fertilizers. Bull 10, Indian Soc Soil Sci, New Delhi, pp 144–163
Hadlow W, Millard EW (1981) Minimum tillage: a practical alternative to ploughing in South African sugar industry. Sugar J (USA) 44: 15–17
Hall OD (1979) Solar energy use through biology – past, present and future. Sol Energy 22: 307–328
Halmann M (1990) Synthetic plant growth regulators. Adv. Agron 43: 47–105
Ham GJ (1982) The effect of ratooning depth on ratoon crop performance. Proc Conf Aust Soc Sugar Technol Townville, pp 155–159
Hamilton DP (1992) The wonders of the neem tree revealed. Science 255: 275
Hapse DG, Deshmukh HS, Gunjul BB, Shinde PP (1990) Evaluation of a microirrigation system in sugarcane agriculture. XI Int Cong on the use of plastics in agriculture, Feb 26 – March 2, 1990, New Delhi, pp 41–47
Harmsen GW (1954) Observations on the formation and oxidation of pyrite in the soil. Plant Soil 5: 324–348
Harrison JB, Bovell JR (1888) Reports on the results obtained upon the experimental fields at Dodds Reformatory, Barbados
Hartt CE (1929) Potassium deficiency in sugarcane. Bot Gaz 88: 229–261
Hartt CE (1959) The growth of sugarcane as influenced by phosphorus – the critical range of plant phosphorus. Int Soc Sugarcane Technol Proc 10: 467–473
Hartt CE (1970) Effect of potassium deficiency upon translocation of ^{14}C in detached blades of sugarcane. Plant Physiol 45: 183–187
Hartt CE (1972) Translocation of carbon-14 in sugarcane supplied with or deprived of phosphorus. Plant Physiol 49: 569–571
Hartt CE, Burr GO (1965) Factors affecting photosynthesis in sugarcane. Int Soc Sugarcane Technol Proc 12: 590–609
Haysom MBC, Chapman LS (1975) Some aspects of calcium silicate trials at Mackay. Queensl Soc Sugarcane Technol Proc 42: 117–122
Hebert LP, Matherne RJ, Davidson LG (1965) Row spacing experiments with sugarcane in Louisiana. Int Soc Sugarcane Technol Proc 12: 96–102
Heinz DJ (1987) Sugarcane improvement through breeding. Development in crop science, II. Elsevier, Amsterdam 603 pp
Heinz DJ, Mee GWP (1969) Plant differentiation from callus tissue of *Saccharum* species. Crop Sci 9: 346–348
Hendre RR, Iyer RS, Kotwal M, Khuspe SS, Mascarenhas AF (1983) Rapid multiplication of sugarcane by tissue culture. Sugarcane 1 (May-June): 5–8
Hilton HW, Orgood RV, Maretzki A (1980) Some aspects of Mon 8000 as a sugarcane ripener to replace polaris. Int Soc Sugarcane Technol Proc 17: 652–661
Hodnett GE (1956) The responses of sugarcane to fertilisers. J Exp Agric 24: 1–19
Hodnett MG, Bell JP, Ah Khoon PD, Soopramanien GC, Batchelor CH (1990) The control of drip irrigation of sugarcane using index tensiometers: some comparisons with control by the water budget method. Agric Water Manage 17: 189–207

Hornby D (1990) Root diseases. In: Lynch JM (ed) The rhizosphere. Wiley, New York, pp 250–251

Hsia YJ, Ou-Yang C (1968) Studies on poor ratooning of sugarcane in Taiwan. Int Soc Sugarcane Technol Proc 13: 623–634

Hudson JC (1968) Available soil water and sugarcane growth and transpiration. Int Soc Sugarcane Technol Proc 13: 484–498

Humbert RP (1959) Soil as a factor in varietal yield decline. Int Soc Sugarcane Technol Proc 10: 51–62

Humbert RP (1962) Potash and sugarcane quality. Int Soc Sugarcane Technol Proc 11: 115–123

Humbert RP (1968) The growing of sugarcane. Elsevier, Amsterdam, 779 pp

Humbert RP (1975) Harvesting the sun with sugarcane. World Farming 17: 18–20

Humbert RP (1976) The ripening of sugarcane. In: Sugarcane Ripener Seminar Proc Florida, May 2–4, 1976, Monsanto Agric Products Co, St Louis, Missouri, pp 2–9

Humbert RP (1978) Potash usage in growing sugarcane for energy. Potash Rev Subject 27, 87: 1–4

Humbert RP, Martin JP (1955) Nutritional deficiency symptoms in sugarcane. Hawaii Plant Rec 55: 95–102

Hume CJ, Cattle H (1990) The green house effect, meteorological mechanisms and models. Outlook Agric 19: 17–23

Hunsigi G (1982a) Modelling fertilizer responses in sugarcane. Deccan Sugar Technol Assoc 32nd Conv, Poona, A69–A76

Hunsigi G (1982b) Ratooning of sugarcane in India. Prabhat, Mysore, 97 pp

Hunsigi G (1985) The production technology for sugar and sugarcane: an innovative approach. Maharashtra Sugar 9, June 1984: 1–17

Hunsigi G (1989) Agricultural fibres for paper pulp. Outlook Agric 18: 96–103

Hunsigi G (1991) Nonwood fibres for paper pulp: sugarcane and some agricultural crops as sources. UAS Tech Ser 42: 159 pp

Hunsigi G, Srivastava SC (1975) An approach toward a plant type concept in sugarcane. Curr Sci 44: 897–899

Hunsigi G, Srivastava SC (1976) A critical review on the validity of chemical extractants for soil potassium with special reference to sugarcane as test crop. Geophytology 6: 259–263

Hunsigi G, Srivastava SC (1977) Modulation of ET values of sugarcane because of high water table. Int Soc Sugarcane Technol Proc 16: 1557–1564

Hunsigi G, Srivastava SC (1980) Quantity-potential (Q/P) and quantity-intensity (Q/I) isotherms of soil potassium in some sugarcane-growing soils. Deccan Sugar Technol Assoc 30th Conv, Poona, pp 63–72

Hunsigi G, Srivastava SC (1981) Some measures of potassium availability to sugarcane. Fert News 26 June 1981: 35–38

Hunsigi G, Srivastava SC (1983) Potassium fertilization of sugarcane. Sugarcane (UK) 1: 9–12, 18

Hunsigi G, Havanagi GV, Puttaraj Urs B (1974a) Response of sugarcane to applied fertilizers. Potash Rev Subject 27, 66: 1–4

Hunsigi G, Nadagoudar BS, Puttaraj Urs B (1974b) Response of sugarcane to phosphate fertilization at graded levels of nitrogen. Indian J Agric Sci 44: 513–516

Hunsigi G, Nair VSK, Laad VV (1986) Unconventional source of raw materials for the paper industry. A search for alternatives. Bagasse and mesta. Int Soc Sugarcane Technol Proc 19: 30 (poster session)

Husz GS (1972) Sugarcane cultivation and fertilization. RUHR-STICKSTOFF AG Bochum, West Germany, 116 pp

Iggo GA, Moberly PK (1976) The concept of minimum tillage in sugarcane. Proc S Afr Sugar Technol Assoc June 1976: 1–3

Ingram KT, Hilton HW (1986) Nitrogen – potassium fertilization and soil moisture effects on growth and development of drip-irrigated sugarcane. Crop Sci 26: 1034–1039

Inman-Bamber NG, Dejager JM (1986) The reaction of two varieties of sugarcane to water stress. Field Crops Res 14: 15–28

Innes RF (1959) The potash manuring of sugarcane. Int Soc Sugarcane Technol Proc 10: 441–450

Innes RF, Chinloy T (1951) The effects of fertilizers on sugarcane. I. Potash. Proc British West Indies Sugarcane Technol 1951: 56–72

Irvine JE (1967) Photosynthesis in sugarcane varieties under field conditions. Crop Sci 7: 297–300

Irvine JE (1968) Effects of an early freeze on Louisiana sugarcane. Int Soc Sugarcane Technol Proc 13: 837–839

Irvine JE (1975) Relations of photosynthetic rates and leaf and canopy characters to sugarcane yield. Crop Sci 15: 671–676

Irvine JE (1977) Sugarcane. In: Meade GP, Chen JCP (eds) Cane Sugar Hand Book, 10th edn. Wiley, New York, pp 3–14

Irvine JE (1983) Sugarcane. In: Symposium on potential productivity of field crops under different environments. Int Rice Res Inst, Manila, pp 361–381

Irvine JE, Benda GTA (1980) Sugarcane spacing. II. Effect of the spacing on the plant. Int Soc Sugarcane Technol Proc 17: 357–366

Irvine JE, Benda GTA, Richard CA (1980a) Sugarcane spacing. I. Historical and theoretical aspects. Int Soc Sugarcane Technol Proc 17: 350–356

Irvine JE, Richard CA, Garrison DD, Jackson WR, Matherne RT, Camp C, Carter C (1980b) Sugarcane spacing. III. Development of production techniques for narrow rows. Int Soc Sugarcane Technol Proc 17: 368–375

Irvine JE, Richard CA, Carter CE, Dunckelman JW (1984) The effect of row spacing and subsurface drainage on sugarcane yields. Sugarcane 2: 3–5

Isobe M (1968) Water utilization. II. Yield-water relationship. Int Soc Sugarcane Technol Proc 13: 49–54

Jaffri SMH (1980) Residual phosphate in Entisol and its effect on sugarcane ratoons. Proc Deccan Sugar Technol Assoc 30th Conv, Poona, pp 45–50

Jagannath Rao E, Narasimha Rao (1990) Transfer of technology and cane development in sugar factory zones. In: Rao PN (ed) Recent advances in sugarcane. KCP Ltd, Vuyyuru, Andhra Pradesh, pp 487–497

Jones CA (1985) C_4 grasses and cereals. Growth, development and stress response. Wiley, New York, 419 pp

Juang PY (1968) A study of flowering behaviour of sugarcane in different geographic conditions. Int Soc Sugarcane Technol Proc 13: 840–848

Juang TC, Uehara G (1971) Effects of ground water table and soil compaction on nutrient element uptake and growth of sugarcane. Int Soc Sugarcane Technol Proc 14: 679–687

Kanwar RS, Kaur H (1977) Improving sprouting of stubble crop in low temperature areas. Int Soc Sugarcane Technol Proc 16: 1325–1331

Kanwar RS, Sharma KK (1981) Polythene bag planting technique for seed propagation and replanting of ratoons. National seminar on ratoon management, Lucknow, India IISR March 1981: 39–45

Kay M (1990) Recent development for improving water management in surface and overhead irrigation. Agric Water Manage 17: 7–23

Khan AK, Zende GK (1977) The site for Zn – P interaction in plants. Plant Soil 46: 259–262

King LT (1966) Weeds of the world, biology and control. Leonard Hill Books, London, 526 pp

Kulkarni DG (1971) Tropical sugar. Dnyan Jyot Prakashan, Poona, 204 pp

Kumaresan KR, Savithri P, Manickam TS, Kothandaraman GV, Danial KC (1985) Comparative efficacy of soil and foliar application of $ZnSO_4$ and $FeSO_4$ on sugarcane. Madras Agric J 72: 190–195

Lakshmikantham M (1983) Technology in sugarcane growing. IBH Publ, New Delhi, 259 pp

Landisch MR, Svarezkopf JA (1991) Ethanol production and cost of fermentable sugars from biomass. Biores Technol 36: 83–95

Leather JW (1910) Water requirements of crops in India. Mem Dep Agric India Chem Ser 10: 205

Leibbrandt UB (1984) The effects of irrigated ratoon cane by ripping the inter-row after harvest in a range of soils in Swaziland. S Afr Sugar Yearb 55: 13–15

Levirington KC, Sedl JM, Burge JR (1962) Some problems in predicting potassium requirements of sugarcane. Int Soc Sugarcane Technol Proc 11: 123–130

Lima E, Boddey RM, Dobereiner J (1987) Quantification of biological nitrogen fixation associated with sugarcane using a ^{15}N-aided nitrogen balance. Soil Biol Biochem 19: 165–170

Lipinsky ES (1978) Fuels from biomass: integration with food and material systems. Science 199: 644–651

Locsin GL, Gullermo RJ, Tabayoyang FT (1956) Potash fertilization tests on sugarcane in the Victoria district (a review). Sugar News 32: 69–84

Loomis RS, Williams WA (1963) Maximum productivity: an estimate. Crop Sci 3: 67–72

Luthin JW (1957) Preface. In: Luthin JW (ed) Drainage of agricultural lands. ASA Monogr V7, Madison, Wisconsin, 620 pp

Lynch JM (1990) Introduction: some consequences of microbial rhizosphere competence for plant and soil. In: Lynch JM (ed) The rhizosphere, Wiley, New York, pp 1–10

Lynd LR, Cushman JH, Nichols RJ, Wyman CE (1991) Fuel ethanol from cellulosic biomass. Science 251: 1318–1323

Maretzki A, Dela Cruz A (1967) Nitrate reductase in sugarcane tissues. Plant Cell Physiol 8: 605–611

Matherne RJ (1972) Influence of inter-row spacing and planting rate on stalk population and cane yield in Louisiana. Int Soc Sugarcane Technol Proc 14: 640–645

May RM (1980) The world's worst weeds. Nature (Lond) 290: 85–86

McCatty T (1980) A review of sucrose enhancer trials in Jamaica in 1974–78. Int Soc Sugarcane Technol Proc 17: 630–643

Mc Intyre G, Barbe C (1989) The effect of sugarcane yield, weed control, physical and chemical properties of the soil of the minimum tillage vs normal mechanical tillage. Int Soc Sugarcane Technol Proc 20: 586–590

Meade GP, Chen JCP (1977) Cane sugar hand book, 10th edn. Wiley, New York, 947 pp

Mehrad B (1968) Effect of soil salinity on sugarcane cultivation at Haft Tappeh, Iran. Int Soc Sugarcane Technol Proc 13: 746–755

Meyer JH (1981) An evaluation of DRIS based on leaf analysis for sugarcane in South Africa. Proc S Afr Sugar Technol Assoc June 1981: 1–8

Meyer E, Soevey TC, Meyer JH (1989) The incorporation of lime under minimum tillage conditions. Proc Annu Cong S Afr Sugar Technol Assoc 1989, 63: 142–146

Millard E (1974) Plastic mulching of sugarcane. Proc S Afr Sugar Technol Assoc 48: 53–57

Miller E (1990) A low head drip irrigation system for small holdings. Agric Water Manage 17: 37–47

Mishra A, Naidu KM, Gupta ML (1990) Effect of row spacing and nitrogen levels on ethanol production in sugarcane at different ages of crop harvest. Indian J Agri Sci 60: 110–114

Moberly PK (1969) The effects on ratoon cane of subsoiling in a number of soils in the sugar belt. Proc S Afr Sugar Technol Assoc June 1969: 1–3

Moberly PK (1972) Deep tillage investigations on five soil types of the South African sugarbelt. Proc S Afr Sugar Technol Assoc June 1972: 1–6

Mongelard JC (1968) The effect of different water regimes on the growth of two sugarcane varieties. Int Soc Sugarcane Technol Proc 13: 643–651

Mongelard JC, Mimura L (1971) Growth studies on the sugarcane plant. Crop Sci 11: 795–800

Monteith NH, Banath CL (1965) The effect of soil strength on sugarcane growth. Trop Agric 42: 293–296

Moore PH (1989) Physiological basis for varietal improvement. In: Naidu KM, Srinivasan TV, Premchandran MN (eds) Proc Int Symp sugarcane varietal improvement. Present status and future thrusts. Sugarcane Breeding Institute Coimbatore to commemorate its Platinum Jubilee. Sept 3–7, 1987. SBI, Coimbatore, pp 19–55

Morelli JL, Dalben AE, Almeida JOC, Dematte JLI (1989) The effects of increasing amounts of lime and phosphogypsum on the productivity of sugarcane and on chemical properties of sandy acid soil. Int Soc Sugarcane Technol Proc 20, 2: 505–511

Mukherjee SK (1957) Origin and distribution *Saccharum*. Bot Gaz 119: 55–61

Naidu KM (1986) Recent strategies for improving sugar and sugarcane productivity. National conference on sugarcane production: prospects and perspectives. Indian Inst Sugarcane Res Lucknow Feb 15–16: 1–32

Naidu KM (1989) Sugarcane varietal improvement. In: Naidu KM, Srinivasan TV, Premchandran MN (eds) Proc Int Symp sugarcane varietal improvement. Present status and future thrusts. Sugarcane Breeding Institute Coimbatore to commemorate its Platinum Jubilee, Sept 3–7, 1987. SBI, Coimbatore, pp 1–17

Narasimhan RL (1957) Fibre in sugarcane. A review. Indian J Sugarcane Res Dev Part I: 31–35

Nemeth K (1975) The effect of K fertilization and K removal by rye grass in pot experiment on the K concentration of the soil solution of various soils. Plant Soil 42: 97–107

Ng Kee Kwong KF, Deville J (1988) Filter cake as a nitrogen fertilizer for plant cane in Mauritius. Trop Agric (Trinidad) 65: 364–366

Nour AH, El-Wakeel AF, Alam AI (1989) Effect of soil salinity on sugarcane yield. Int Soc Sugarcane Technol Proc 20, 2: 532–535

Novoa R, Loomis RS (1981) Nitrogen and plant production. Plant Soil 58: 177–204

Orgill B (1987) Newsprint from bagasse and hardwood pulp. Co-op Sugar 18: 397–407

Ormord DP, Black VJ, Unsworth MH (1981) Depression of net photosynthesis in *Vicia faba* exposed to sulphur dioxide and ozone. Nature (Lond) 291: 585–586

Panje RR (1970) Observations on the propagation and commercial application of the new sugarcane planting method 86206. Indian Sugar 19, 10: 1–4

Panje RR, Srinivasan K (1959) Studies in *Saccharum spontaneum*. The flowering behaviour of latitudinally displaced populations. Bot Gaz 120: 193–202

Parish DH (1965) The efficacy of nitrogenous fertilizers. Int Soc Sugarcane Technol Proc 12: 372–379

Parthasarathy N (1948) Origin of noble sugarcane (*Saccharum officinarum* L). Nature (Lond) 161: 608

Parthasarathy SV (1972) Sugarcane in India. KCP Ltd Madras 804 pp

Paturau JM (1986) The state of the art in the utilization of by-products of the cane industry. Key Note Pap Int Soc Sugarcane Technol Proc 19: IV, 1–IV 16

Paturau JM (1989) By-products of the cane sugar industry: an introduction to their industrial utilization. Sugar series 11. Elsevier, Amsterdam, 389 pp

Payne D (1988) Soil structure, tilth and mechanical behaviour. In: Wild A (ed) Russell's soil conditions and plant growth, 11th edn. ELBS Longman, Great Britain, pp 378–411

Payne D, Gregory PJ (1988) The soil atmosphere. In: Wild A (ed) Russell's soil conditions and plant growth, 11th edn. ELBS Longman, Great Britain, pp 298–314

Peng SY (1984) The biology and control of weeds. In: Sugarcane: developments in crop science (4). Elsevier, New York, 366 pp

Penman HL (1948) Natural evapotranspiration from open water, bare soil and grass. Proc R Soc Lond 193: 120–145

Pesek JT (1974) Crop yield response equations and economic levels of fertilizer use. In: Fernandez HV (ed) Fertilizers, crop quality and economy. Elsevier, Amsterdam, pp 881–933

Plucknett DL, Evenson JP, Stanford WG (1970) Ratoon cropping. Adv Agron 22: 285–330

Prasad M (1976) Response of sugarcane to filter press mud and NPK fertilizers. I. Effect on sugarcane yield and sucrose content. Agron J 68: 539–543

Prasad SR, Sharma RK, Alam M (1981) Yield of ratoon in relation to intensity of gaps. National seminar on ratoon management, vol 1. Indian Inst Sugarcane Res Lucknow: 72–74

Pruett RL (1981) Synthesis gas: a raw material for industrial chemicals. Science 211: 11–16

Qu-Ming Lin, Ju-Ming Chen (1986) Status and management of magnesium on the upland red soils of Fujian China. Proc Int Conf on management and fertilization of upland soils in the tropics and subtropics. Sept 7–11, 1986. Nanjing, People's Republic of China, China Consultants Int (HK) Ltd, pp 229–230

Rajan BCK (1987) Versatile *Eucalyptus*. Diana, Bangalore, 220 pp

Ramalingashetty TK (1986) Bagasse-based paper mills. Some Indian experiences and financial aspects. IPPTA Bangalore, 22: 62–76

Rangamannar G, Ramaswamy K (1986) Biotechnical aspects of various bagasse storage systems. IPPTA Bangalore Conf, 22: 1–15

Rao JT (1989) Sugarcane origin, taxonomy, breeding and varieties. In: Naidu KM, Srinivasan TV, Premchandran MN (eds) Proc Int Symp on sugarcane varietal improvement. Present status and future thrusts. Sugarcane Breeding Institute, Coimbatore, to commemorate its Platinum Jubilee, Sept 3–7, 1987, SBI, Coimbatore, pp 83–113

Rau S, Millard EW (1975) Further studies on the use of polyethylene mulch in the growing of sugarcane. Proc S Afr Sugar Technol Assoc June/July 1975: 1–5

Reid CPP (1990) Mycorrhizas. In: Lynch JM (ed) The rhizosphere. Wiley, New York, pp 281–315

Reynolds WN, Gibson W (1968) Water utilization. III. Sugarcane irrigation in Hawaii. Int Soc Sugarcane Technol Proc 13: 55–67

Ricaud R (1971) Effects of varying numbers of cultivations on sugarcane production in Louisiana. Int Soc Sugarcane Technol Proc 14: 634–639

Ricaud, R (1977) Effect of sub-soiling on soil compaction and yield of sugarcane. Int Soc Sugarcane Technol Proc 16: 1039–1047

Ricaud R, Arceneaux A (1986) Some factors affecting ratoon cane yield and longevity in Louisiana. Int Soc Sugarcane Technol Proc 19, 1: 18–24

Richards SJ, Hagan RM, McCalla JM (1952) Soil temperature and plant growth. In: Shaw BT (ed) Soil physical conditions and plant growth. Academic press, New York, pp 303–480

Roach BT, Daniels J (1987) A review of the origin and improvement of sugarcane. Copersucar Inter Sugarcane Breeding Workshop, May–June 1987, pp 1–31

Roberts J, Nayamuth RA, Batchelor CH, Soopramanien GC (1990) Plant – water relations of sugarcane under a range of irrigated treatments. Agric Water Manage 17: 95–115

Rochecouste E (1959) Evaluating herbicides in sugarcane cultivation. Int Soc Sugarcane Technol Proc 10: 549–555

Rogers PL, Lee KJ, Tribe DE (1983) High productivity of ethanol fermentation with *Zymomonas mobilis*. Process Biochem Aug–Sept 1980: 7–11

Rosillo-Calle F, Rothman H (1986) The Brazilian National Alcohol Programme: some technological and socio-economic aspects. Outlook Agric 15: 196–201

Rowell DL (1988) Soil acidity and alkalinity. In: Wild A (ed) Russell's soil conditions and plant growth. ELBS, Longman, Great Britain, pp 844–898

Ruschel AP, Ruschel R (1977) Varietal differences affecting nitrogenase activity in the rhizosphere of sugarcane. Int Soc Sugarcane Technol Proc 16: 1941–1947

Rydholm SA (1965) Pulping process. Inter Sci, New York, 692 pp

Sahi BK, Sundara B (1986) Short duration sugarcane. Kissan World Aug 1986: 35–37

Samuels G (1969a) Silicon and sugar. Sugar Y Azucar 64: 25–29

Samuels G (1969b) Foliar diagnosis for sugarcane. Adams press, Chicago, 362 pp

Samuels G, Alexander AG (1968) Influence of variable manganese and silicon on the nutrition, sugar production and enzyme activity of immature sugarcane. Int Soc Sugarcane Technol Proc 13: 545–555

Samuels G, Landrau Jr P (1956) The sucrose content of sugarcane as influenced by fertilizers. Int Soc Sugarcane Technol Proc 9: 365–375

Sanchez PA (1976) Properties and management of soils in the tropics. Wiley, New York, 618 pp

Saroha MS, Singh HG (1980) Interrelationship between iron and sulphur in the prevention of chlorosis in sugarcane in alkaline calcareous soils. Indian J Agric Sci 50: 34–40

Sastry KNR, Ramanna R (1978) Economics of growing planted and ratoon crops of sugarcane in Krishnaraj Sagar area. Financing Agric Jan–March, 1978: 10–13

Sedgley RH (1991) An appraisal of the Donald ideotype after 21 years. Field Crops Res 26: 93–112

Sharma MP, Singh K (1985) Partial mechanization of sugarcane harvesting. Agric Mech Asia, Africa Latin Am 16: 347–350

Sharma MP, Singh K (1988) IISR implements and machinery for sugar crops. Tech Bull No 20 Indian Institute of Sugarcane Research Lucknow, 63 pp

Shew MEA (1989) An index to measure sugarcane ratoon performance. Sugar Y Azucar 84(7): 19, 22–2326

Shih SF (1985) Use of sugarcane yield and evapotranspiration relationships to improve water management. Sugarcane 1: 3–7

Shih SF (1989) Relating calculated leaf area index, evapotranspiration and irrigation method of sugarcane. Agron J 81: 111–115

Shih SF, Gascho GJ (1980) Relationship among stalk length, leaf area and dry biomass of sugarcane. Agron J 72: 309–313

Shoji K (1965) Weed control in Puerto Rico. Int Soc Sugarcane Technol Proc 12: 737–741

Shoji K (1977) Drip irrigation. Sci Am 237: 62–68

Shoji K, Sund KA (1965) Drainage and salinity investigation at the Haft Tappeh sugarcane project, Iran. Int Soc Sugarcane Technol Proc 12: 90–95

Shrivastava AK, Singh K, Kacker NK (1981) Partitioning of dry matter, energy conversion efficiency and energy productivity in plant cane and ratoons. National seminar on ratoon management, vol 1. Indian Inst Sugarcane Res Lucknow: 82–91

Shrivastava AK, Singh K, Saxena YR (1985) Production physiology of sugarcane cultivar CO1148. Tech Bull No 15. Indian Inst Sugarcane Res, Lucknow, 106 pp

Silva deA GM, Degaspari N, Carvalho LCC, Rodrigues de PA, Baglione BM (1989) The use of ethephon to manage sugarcane varieties in different locations of the central-southern region of Brazil. Int Soc Sugarcane Technol Proc 20: 635–646

Simmonds NW (1986) Sugarcanes. In: Simmonds NW (ed) Evolution of crop plants. Longmans, London, pp 104–106

Singh K (1981) Preamble to the national seminar on ratoon management vol I. Indian Inst Sugarcane Res, Lucknow, March 1981: 1–2

Sinha OK (1992) Tissue culture in sugarcane improvement. Sugar Crops Newslett 2(1): 4–6

Sinha SK, Swaminathan MS (1979) The absolute maximum food production potential in India: an estimate. Curr Sci 48: 425–429

Soil Survey Staff (1975) Soil taxonomy. US Dept Agric Hand Book 436, Washington DC, 754 pp

Soltwedel F (1986) Verlag der WerkZaamheden en het proefstation te Semarang, gedurende de maand Mei 1886 Tifschr, Land en Tuinb en Boschcult 2, 104

Soopramanien GC, Nayamuth R, Batchelor CH (1989) Effect of water regime on the yield of drip-irrigated sugarcane intercropped with maize. Int Soc Sugarcane Technol Proc 20, 2: 537–543

Soopramanien GC, Teeluck M, Mestry J (1990) Ripener effect on cane growth, sucrose content and regrowth. Sugarcane (UK) Spring Suppl: 1–4

Srinivasan KV (1968) The role of the rhizosphere microflora in the resistance of sugarcane to *Pythium* root rot. Int Soc Sugarcane Technol Proc 13: 1224–1236

Srinivasan TR (1989) Varietal responses to climate, population dynamics, nutrition and other inputs. In: Naidu KM, Srinivasan TV, Premchandran MN (eds) Proc Int Symp on sugarcane varietal improvement. Present status and future thrusts. Sugarcane Breeding Institute, Coimbatore, to commemorate its Platinum Jubilee, SBI, Coimbatore, pp 195–220

Srinivasan TR, Naidu KM (1987) Response of sugarcane varieties to biofertilizers under different soil conditions. Sugarcane (UK) 3: 5–11

Srivastava SC (1966) The effect of telodrin on nitrification of ammonia in soil and its implication on nitrogen nutrition of sugarcane. Plant Soil 25: 471–473

Srivastava SC, Johari DP (1979) The irrigated sugarcane in India. ICAR Indian Inst Sugarcane Res, Lucknow, 75 pp

Stanford G (1973) Rationale for optimum nitrogen fertilization and corn production. J Environ Qual 2: 159–166

Stanford G, Ayres AS (1964) The internal nitrogen requirement of sugarcane. Soil Sci 98: 338–344

Stanford G, Jordan HV (1966) Sulphur requirements of sugar, fiber and oil crops. Soil Sci 101: 258–266

Stevenson GC (1965) Genetics and breeding of sugarcane. Longman's Green and Co, London, 34 pp

Stewart MJ (1969) Potassium and sugarcane. South African Sugar Industry Agronomists Assoc, Rev Pap 5, 15 pp

Stolzy LH, Jury WA (1982) Soil as a physical system. In: Kilmer VJ (ed) Hand book of soils and climate in agriculture. CRC series in agriculture, Boca Raton, pp 131–158
Stone R (1992) A biopesticidal tree begins to bloom. Science 255: 1070–1071
Sundara B, Subramanian S (1987) Energy input-output studies of some crop sequences based on short duration sugarcane in relation to the conventional sugarcane system in tropical India. Agric Ecosyst Environ 20: 49–57
Sundara B, Subramanian S (1989) Fitting short duration sugarcane into cropping systems: yield and quality of short duration sugarcane as influenced by row spacings and nitrogen levels in different sequences. Madras Agric J 76: 129–135
Tabayoyong FT (1959) The effects of soil types and fertilization on the yield of cane in the Philippines. Int Soc Sugarcane Technol Proc 10: 398–408
Tabayoyong FT, Robinol M (1962) Yield response of sugarcane to fertilizers. Int Soc Sugarcane Technol Proc 11: 177–185
Talibudeen O (1974) The nutrient potential of the soil. Soils Fertil 37: 41–45
Tandon HLS (1991) Sulphur research and agricultural production in India, 3rd edn. The Sulphur Institute, Washington DC, 140 pp
Tang KH, Ho FW (1968) A study of nine consecutive ratoons and various methods of manipulating fertility in Taiwan. Int Soc Sugarcane Technol Proc 13: 618–622
Thomas JR, Scott Jr AW (1990) Effect of nitrogen fertilization on availability of P and K to sugarcane. Sugarcane (UK) 2: 10–14
Thompson GD (1965) The relationship of potential evapotranspiration of sugarcane to environmental factors. Int Soc Sugarcane Technol Proc 12: 3–9
Thompson GD (1966) Trash mulching – its effects on soils and crops in South Africa. World Crops June 1966: 62–65
Thompson GD (1976) Water use by sugarcane. Rev Pap 8, S Afr Sugar J 60: 593–635
Thompson GD, du Toit JL (1965) The effects of row spacings on sugarcane crop in Natal. Int Soc Sugarcane Technol Proc 12: 103–112
Thompson GD, Pearson CHO, Cleasby TG (1963) The estimation of the water requirements of sugarcane in Natal. Proc Annu Cong S Afr Sugar Technol Assoc Proc 37: 134–142
Thornley JHM, Johnson IR (1990) Plant and crop modelling: a mathematical approach to plant and crop physiology. Oxford Sci Publ, p 324
Thornthwaite CW (1948) An approach toward a rational classification of climate. Geogr Rev 38: 55–94
Tisdale SL, Nelson WL, Beaton JD (1988) Soil fertility and fertilizers, 4th edn. Macmillan, New York, 754 pp
Trenbath BR (1986) Resource use by intercrops. In: Francis CA (ed) Multiple cropping systems. Macmillan, New York, pp 57–81
Trouse AC Jr, Humbert RP (1961) Some effects of soil compaction on the development of sugarcane roots. Soil Sci 91: 208–217
True RH (1922) Significance of calcium for higher green plants. Science 1955: 1–6
Tudge C (1988) Food crops for the future. Basil Blackwell, New York, 229 pp
Turner NC (1990) Plant water relations and irrigation management. Agric Water Manage 17: 59–73
US Salinity Laboratory Staff (1954) Diagnosis and improvement of saline and alkali soils. Agric Hand Book 60, US Dep of Agriculture, Washington DC, 160 pp
Vandervorm PDJ (1980) Uptake of Si by five plant species as influenced by variations in Si supply. Plant Soil 56: 153–156
Van Dillewijn C (1952) Botany of sugarcane. Chronica Botanica Co, Waltham, MA, 371 pp
Veeresh GK (1977) Studies on the root grubs of Karnataka. UAS Monogr 2, 87 pp
Venkataramana S, Shanmugasundaram S, Naidu KM (1984) Growth behaviour of field-grown sugarcane varieties in relation to environmental parameters and soil moisture stress. Agric For Meteor 31: 251–260
Venkatraman TS (1937) Sugarcane bamboo hybrids. Indian J Agric Sci 7: 513–514
Viets FG Jr (1962) Fertilizers and the efficient use of water. Adv Agron 14: 233–264
Villamil G, Rodriguez L, Bambanaste R (1986) Optimised chemi-mechanical bagasse pulp for newsprint. Int Soc Sugarcane Technol 19 (Abstr): 22
Vlitos AJ (1974) A review of plant growth regulating chemicals in sugarcane cultivation. Proc Int Soc Sugarcane Technol (plant physiol) 15: 932–937
Vomacil JA (1957) Measurement of soil bulk density and penetrability: a review of methods. Adv Agron 9: 159–175

Von Uexkull HR (1968) Potassium nutrition of tropical crops. In: Kilmer VJ, Younts S, Brady MC (eds) The role of potassium in agriculture. Am Soc Agron Crop and Soil Sci Soc Am Wisconsin: 385–421

Wang PL, Yang PC, Chang YT, Yao TP (1987) Effects of subsurface changes on ratoon growth and yield in medium textured soils. Rep Taiwan Sugar Res Inst 116: 7–17

Weissinger AK (1992) Physical methods for plant gene transfer. In: Moss JP (ed) Biotechnology and crop improvement in Asia. Int Crop Res Inst, Semi Arid Tropics (ICRISAT), Patancheru, Andhra Pradesh, India, pp 213–233

Wild A (1988a) Plant nutrients in soil. In: Wild A (ed) Russell's soil conditions and plant growth, 11th edn. Longman, Great Britain, pp 652–779

Wild A (1988b) Potassium, sodium, calcium, magnesium, sulphur, silicon. In: Wild A (ed) Russell's soil conditions and plant growth, 11th edn. Longman, Great Britain, pp 743–779

Willey RW (1979) Intercropping – its implications and research needs. Part I. Competition and yield advantages. Field Crop Abstr 32: 1–10

Wong You Cheong Y, Halais P (1976) Needs of sugarcane for silicon when growing in highly weathered latosols. Trop Agric 6: 99–106

Wood RA (1989) Sugarcane nutrition and fertilizer use in South Africa. Int Soc Sugarcane Technol Proc 20, 2: 495–504

Woodruff CM (1955) The energies of replacement of calcium by potassium in soils. Soil Sci Soc Am Proc 19: 167–171

Wynnparry D, Kelso M (1977) The ultrastructure and analytical microscopy of silicon deposits in the roots of *Saccharum officinarum* L. Ann Bot 44: 855–862

Yadava RL (1991) Sugarcane production technology. Constraints and potentialities. IBH, New Delhi, 291 pp

Yates RA, Taylor RD (1988) Water-use efficiencies in relation to sugarcane yields. Sugarcane 1: 6–10

Zadrazil H (1990) Dragline irrigation: practical experience with sugarcane. Agric Water Manage 17: 25–35

Zende GK (1981) Seminar in soil management for economic production of sugarcane. 31st Annu Conv Deccan Sugar Technol Assoc Proc, Poona, pp S1–S96

Zende GK (1990) Soil fertility management for higher sugar and sugarcane production. In: Rao PN (ed) Recent advances in sugarcane, KCP Ltd, Vuyyuru, Andhra Pradesh, pp 99–200

Zuberer DA (1990) Soil and rhizosphere aspects of N_2 – fixing plant microbe associations. In: Lynch JM (ed) The rhizosphere. Wiley, New York, pp 317–353

Subject Index

Printing: Mercedesdruck, Berlin
Binding: Buchbinderei Lüderitz & Bauer, Berlin